SpringerBriefs in Earth Sciences

More information about this series at http://www.springer.com/series/8897

Philipp Schmidt-Thomé
Thi Ha Nguyen · Thanh Long Pham
Jaana Jarva · Kristiina Nuottimäki

Climate Change Adaptation Measures in Vietnam

Development and Implementation

 Springer

Philipp Schmidt-Thomé
Jaana Jarva
Kristiina Nuottimäki
Geological Survey of Finland (GTK)
Espoo
Finland

Thi Ha Nguyen
National Centre for Water Resources
    Planning and Investigation (NAWAPI)
Hanoi
Vietnam

Thanh Long Pham
Sub-Institute of HydroMeteorology
    and Environment of South Vietnam
    (SYHYMETE)
Ho Chi Minh City
Vietnam

ISSN 2191-5369   ISSN 2191-5377   (electronic)
ISBN 978-3-319-12345-5   ISBN 978-3-319-12346-2   (eBook)
DOI 10.1007/978-3-319-12346-2

Library of Congress Control Number: 2014953247

Springer Cham Heidelberg New York Dordrecht London

Printed on acid-free paper

Springer is part of Springer Science+Business Media (www.springer.com)

Contents

Abbreviations

BOD	Biochemical Oxygen Demand
BOD$_5$	Biochemical oxygen demand over a 5-day period
CC	Climate Change
CEWAFO	Centre for Water Resources Monitoring and Forecast
COD	Chemical Oxygen Demand
DO	Dissolved Oxygen
DONRE	Department of Natural Resources and Environment of Vietnam
Drawdown	Change in groundwater level due to withdrawal
Flow (q)	A volumteric flow rate water discharges from a source (L^3/s)
Flow rate	The rate or speed groundwater moves (m/d)
GDP	Gross Domestic Product
GTK	Geological Survey of Finland
Hydraulic conductivity (K)	A measure of the permeability of the material through which the water is following (m/d)
ICI	Institutional Cooperation Instrument (Ministry for Foreign Affairs of Finland)
IMHEN	Vietnam Institute of Meteorology, Hydrology and Environment
IPCC	International Panel on Climate Change
IZ	Industrial zone
L	Litre, an alterantive symbol for l (lowercase letter L)
MKD	Mekong River Delta
MONRE	Ministry of Natural Resources and Environment of Vietnam
NAWAPI	Centre for Water Resources Planning and Investigation
NTP-RCC	National Target Program of Vietnam to Respond to Climate Change
QCVN	Vietnam Technical Regulations

SEADPRI	Southeast Asia Disaster Prevention Research Institute
SIHYMETE	The Sub-Institute of HydroMeteorology and Environment of South Vietnam
SLR	Sea Level Rise
Specific capacity (Q/s)	The rate of discharge of water from the well divided by the resulting drawdown on the water level within the well (L/s/m)
Specific discharge (q)	The discharge per unit areas normal to flow (L/s)
Specific storage (S_s)	The volume of water released per unit volume of aquifer for a unit decrease in hydraulic head (L^{-1})
Specific yield (S_y)	The volume of water that a saturated medium can yield by gravity drainage per unit volume of the porous medium (dimensionless)
Storage coefficient	cf. Storativity
Storativity (S)	The volume of water released from storage with respect to the change in water level and surface area of the aquifer (dimensionless)
SYKE	Finnish Environment Institute
TDS	Total Dissolved Solids
Transmissivity (T)	The ability of an aquifer to transmit water throughout its entire saturated zone (m^2/d)
TSS	Total Suspended Solids
Yield (Q)	The volume of water pumped or discharged from a well (L/s) or (m^3/d)

References

British Geological Survey (2014) Glossary of groundwater and groundwater-related terms. British Geological Survey. http://www.bgs.ac.uk/research/groundwater/resources/glossary. html. Accessed 13 June 2014

Goulburn-Murray Water (2010) Groundwater—Terms and definitios. Document 2977263. http://www.g-mwater.com.au/downloads/Groundwater/2977263-v5-GROUNDWATER_TERMS_AND_DEFINITIONS_GLOSS-1.PDF. Accessed 13 June 2014

Sharp J M Jr (2007) A glossary of hydrogeological terms. Department of Geological Sciences, The University of Texas, Austin, Texas, 63 pp. http://www.geo.utexas.edu/faculty/jmsharp/sharp-glossary.pdf. Accessed 13 June 2014

Author Biographies

Philipp Schmidt-Thomé is a geographer (M.Sc.) and holds a Ph.D. in Geology. As a senior scientist at the Geological Survey of Finland (GTK), he is specialized in environmental geology, natural hazards, climate change adaptation and land-use planning. He works as a project manager in European research and regional development projects, as well as in international research and consulting and conducts expert assignments by the European Commission and other institutions. He is an adjunct professor to the University of Helsinki and a regular lecturer in several European universities. He leads the Working Group on Climate Change Adaptation of the Commission on Geoscience for Environmental Management of the International Union of Geological Sciences (IUGS-GEM) and is a visiting fellow to the South-East Asia Disaster Prevention Institute (SEADPRI).

Thi Ha Nguyen is an internationally trained hydrogeologist (M.Sc.) and holds a Ph.D. in Hydrogeochemistry. She is a director at the Centre for Water Resources Monitoring and Forecast (CEWAFO), which is one unit of the National Centre for Water Resources Planning and Investigation (NAWAPI). Her working areas cover monitoring, planning, communicating and forecasting water resources. She is also working as a coordinator in science and technology projects, as well as in international assignments.

Thanh Long Pham is a biochemist (M.Sc.) working as a researcher at the Sub-Institute of HydroMeteorology and Environment of South Vietnam (SIHYMETE), which is one unit of the Vietnam Institute of Meteorology, Hydrology and Environment (IMHEN). His research areas cover environment and sustainable development, climate change impacts assessment and adaptation. He is also responsible for scientific and international cooperation at his agency. He has also been working as a coordinator in science and technology projects, as well as in international projects. He is currently preparing his Ph.D. in environment management codes in the field of climate change.

Jaana Jarva is an environmental geologist (M.Sc.) at the Geological Survey of Finland (GTK). She is specialized in the fields of environmental geology, geochemistry, land-use planning, geological risk assessment and climate change adaptation. She has worked as a project manager in several European research and regional development projects, as well as in national and international research and consulting assignments. She is currently finalizing her Ph.D. thesis on urban geochemistry at the University of Turku.

Kristiina Nuottimäki is a geologist (M.Sc.) at the Geological Survey of Finland (GTK). She has vast work experience in contaminated soil remediation and risk assessment in the Nordic countries. She is specialized in environmental geology, GIS, climate change adaptation and land-use planning. At GTK, she is working in climate change adaptation and land-use planning-related projects, as well as in international consulting.

Chapter 1
Preface

Abstract The development of climate change adaptation strategies is nearly state-of-the-art in many countries, but often there is still a large step towards implementing climate change adaptation measures on the local level. The challenge in the development of adaptation measures lies in their acceptability by local stakeholders and decision makers. They also usually demand investments, most of which have to be financed by local municipalities. The uncertainties of climate change models make it difficult to justify investments to finance protection from uncertain impacts. Setting out with the projected climate change impacts in Vietnam, which is one of the most vulnerable countries to climate change, the book then describes a methodological approach to assess and evaluate local vulnerabilities of natural resources to climate change and socio-cconomic impacts, engaging local stakeholders in the development of locally acceptable and economically feasible adaptation measures. Scenario workshops support the communication between scientists and stakeholders. To understand potential future risks the communication methodology was to first get a good understanding of the natural resources (mainly surface and groundwater) and their potential vulnerabilities (current and future). This was followed by developing a common understanding of current risk patterns, as well as underlying vulnerabilities and hazards. Socioeconomic developments have an equally strong, and in the short term mostly even stronger, impact on the living environment and natural resources as long-term climate change impacts. The scenario workshops developed a holistic approach on current and potential future risk patterns, with a special focus on surface and groundwater quantities and qualities, natural hazards and sea level rise. Land-use planning is addressed as playing a decisive role in minimizing current and future risks. The results achieved are applicable to other areas in Vietnam, following the same methodology to identify local solutions with the support of scenario workshops and local stakeholder engagement. The approach is also very well applicable to other countries, certainly respecting different cultural settings in the design of the scenario workshops.

© The Author(s) 2015

P. Schmidt-Thomé et al., *Climate Change Adaptation Measures in Vietnam*, SpringerBriefs in Earth Sciences, DOI 10.1007/978-3-319-12346-2_1

This book describes how climate change adaptation measures were developed for two case study areas in Vietnam. The development and the implementation of climate change adaptation measures is very demanding and complicated because it involves several scientific disciplines, stakeholders and decision makers. One main issue of crucial importance is the decision making to prevent potential risks in the future, where both the climatic conditions, as well as the socio-economic settings of this climate are based on models. Both climate change models and socio-economic development forecasts bare substantial uncertainties, so that adaptation measures, that often demand a shift in thinking as well as investments, need to be justified by the decision makers towards the beneficiaries, as well as the tax payers and investors.

In this climate of uncertainty it is advisable to identify adaptation measures that go along with investments that need to be done anyhow (e.g. infrastructure improvements), and are of societal and business advantage from the very beginning (e.g. clean drinking water). In order to identify potential future vulnerabilities it is thus vital to understand current vulnerabilities first, and then extrapolate these into future development scenarios. These scenarios need to be based on sound scientific understanding of the current living-environment.

This book describes the development of climate change adaptation measures in two case studies in Vietnam, Thanh Hoa and Ba Ria–Vung Tau, from an early stage on. The involved local scientists and local stakeholders identified water quality and quantity as the crucial vulnerabilities of these case studies, followed by natural hazards. The underlying project that developed these adaptation measures, "Development and implementation of climate change adaptation measures in coastal areas in Vietnam—VIETADAPT" (2011–2013) therefore started with a thorough investigation of the ground and surface water conditions of these two case studies, as well as natural-hazard assessments. The results of the scientific research where then analyzed with local stakeholders. Potential future developments and vulnerabilities were jointly assessed in scenario workshops. The results of the scenario workshops were then used to draft and prioritize adaptation measures.

This book starts off with a description of the Vietnamese climate and potential climate change impacts. It then describes in detail the groundwater and surface water field measurements conducted in the two case study areas. The results of the field measures are then overlain with development scenarios in order to determine future vulnerabilities and risks to both water quantity and quality. This is followed by the background theory and the conduction of the scenario workshops. Finally, the adaptation measures, jointly developed by local scientists and local stakeholders are presented.

The VIETADAPT project has been successful in approaching the local communities because it has involved the stakeholders and decision makers in all project activities. The project has developed trust and confidence among local stakeholders, particularly the Department of Natural Resources and Environment (DONRE) and other local management authorities. The local stakeholders reckoned that the project had contributed to the development of awareness about fundamental issues concerning the environment and groundwater resources; as well as the related vulnerabilities and risk caused by both, socio-economic development and climate change impacts. The local stakeholders have expressed their deep satisfaction with the achieved results. The Ministry of Natural Resources (MONRE) stated that this successfully developed methodology shall be applied to identify vulnerabilities to climate change and respective adaptation measures in other provinces of Vietnam.

1.1 Background and Scope

Vietnam is identified by the International Panel on Climate Change (IPCC) as one of the countries likely to be most affected by climate change (IPCC 1994, 2001), especially by sea level rise, changing precipitation and typhoon patterns, and resulting changes in flood patterns, among others. Recognizing potential impacts of climate change, the Government of Vietnam has approved the National Target Program to Respond to Climate Change (2008) (NTP-RCC), which stated that *"climate change response activities must be timely carried out together with their long-term impacts to ensure sustainable development"*. The NTP-RCC led to the Vietnamese "National Action Plan to Respond to Climate Change 2012–2020" (Vietnam Climate Change Adaptation Strategy 2011) which was addressed by the project by developing climate change adaptation measures at the local level. The VIETADAPT project had the overall objective to *"contribute to the establishment of the Vietnamese Climate Change Adaptation Strategy by developing local level planning procedures and adaptation measures"*. The project contributed to the development and implementation of the Vietnamese "National Climate Change Strategy" by deriving applicable local climate change adaptation measures that serve as a model for other areas in Vietnam.

Vietnamese coastal areas are experiencing strong socio-economic development. Both socio-economic development and climate change impacts severely affect human and ecosystems in coastal areas. Sea level rise and flood tides are of great concern for causing salinization of groundwater, surface water, and soil. Therefore, it is important to distinguish between long-term climate change impacts and human socio-economic development impacts. The latter of which also leads to the exploitation of natural resources. It is obvious that both impacts are parallel, but the necessary adaptation measures can be developed and implemented only at the local level, taking local conditions into account. Thus, local stakeholders and decision makers should become aware of using local natural resources in a sustainable manner, taking local climate change impacts, as well as socio-economic structures into account. In the VIETADAPT project, interdisciplinary cooperation between scientists and stakeholders supported the development of sustainable adaptation solutions in case study areas.

The head offices of two Vietnamese partners of the VIETADAPT project, IMHEN and NAWAPI play a crucial role in national climate change adaptation. IMHEN has led various projects involving climate change adaptation and was assigned by the Ministry of Natural Resources and Environment and (MONRE) to act as the focal point in drafting the National Target Program to Respond to Climate Change (NTP-RCC) and to take the lead in implementing the climate change scenarios for Vietnam. Most recently, with the support of the United Nations Development Programme (UNDP), IMHEN successfully completed the Capacity Building for Climate Change project in Vietnam in order to mitigate the impacts of climate change and to control greenhouse gas emissions. NAWAPI has led and conducted various national projects related to water resource surveying and planning, particularly groundwater studies.

GTK has vast experience on leading a number of projects on climate change adaptation in the Baltic Sea Region (SEAREG,[1] ASTRA,[2] BaltCICA[3]), see also Schmidt-Thomé and Klein (2013), as well as on Finnish-Russian cooperation on environmental geology and geological risk assessment for local plans in the city of St. Petersburg, Russia (GeoInforM,[4] CliPLivE,[5] Klein et al. 2013). All these projects were carried out in close cooperation with local strategists and other stakeholders, and have been successful in developing and implementing adaptation solutions. The projects directly solve the core problems in order to adapt to climate change and minimize the impacts of natural hazards and climate change at the local level, addressing the demands set out by UNDP (2004), European Union's White Paper on Adapting to Climate Change (COM 2009), European Union's Strategy on Climate Change Adaptation (COM 2013), IPCC's 4th report (IPCC 2007), and IPCC's 5th report (IPCC 2014) as well as IPCC's special report on extreme phenomena (IPCC 2012).

The VIETADAPT project was financed by the Ministry for Foreign Affairs of Finland under the Institutional Cooperation Instrument (ICI Fund) with the purposes of studying and implementing climate change adaptation solutions via local workshops in two study areas in Vietnam, Thanh Hoa and Ba Ria–Vung Tau (Fig. 1.1). Both case study areas of the project have long coastlines and are currently facing significant socio-economic development in different development stages. Ba Ria–Vung Tau has experienced strong development since several

---

[1] Sea Level Change Affecting the Spatial Development of the Baltic Sea Region—http://www.gtk.fi/projects/seareg.

[2] Developing Policies and Adaptation Strategies to Climate Change in the Baltic Sea Region— http://www.astra-project.org.

[3] Climate Change: Impacts, Costs and Adaptation in the Baltic Sea Region—http://www.baltcica.org.

[4] Integrating Geological Information in City Management to Prevent Environmental Risks— http://www.infoeco.ru/geoinform.

[5] Climate Proof Living Environment—http://www.infoeco.ru/cliplive.

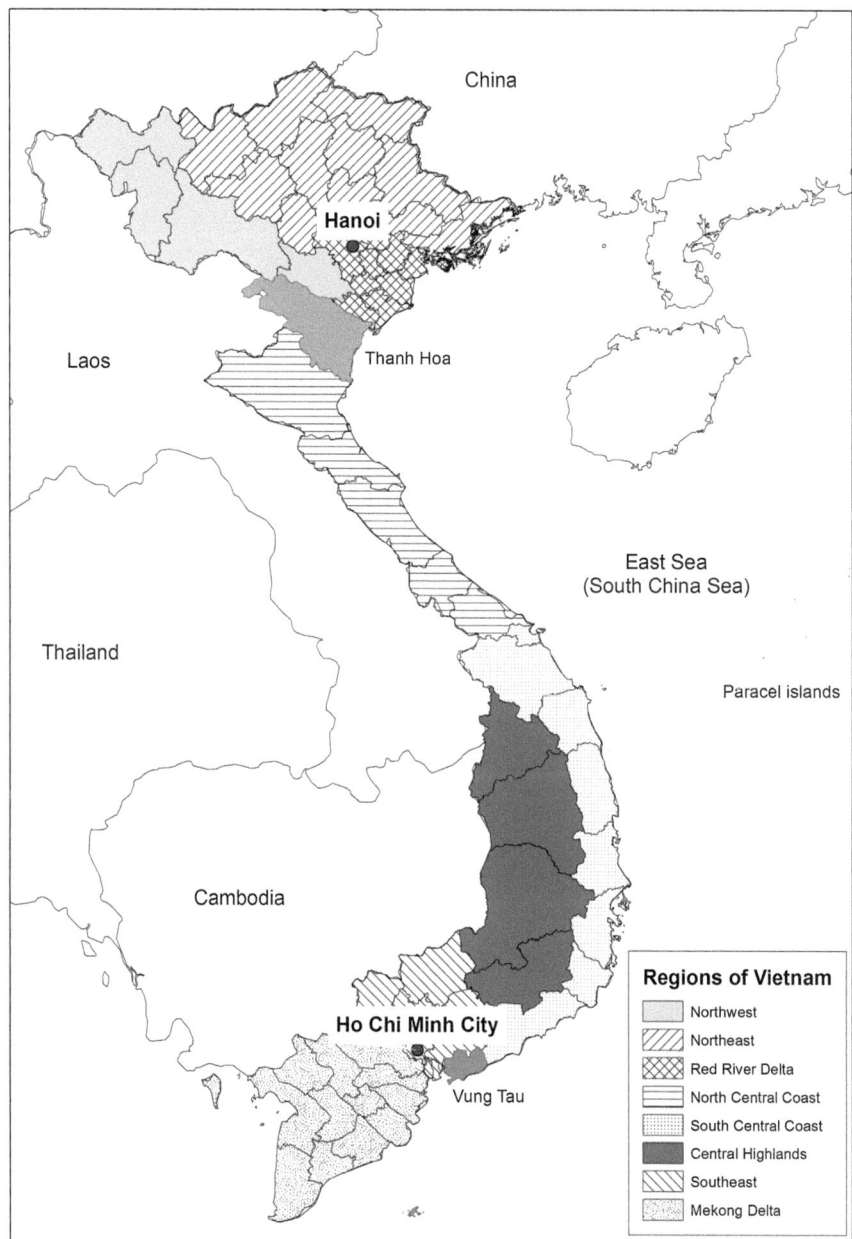

Fig. 1.1 Regions of Vietnam and location of the VIETADAPT project case study areas.
Source ESRI basemap

decades, while Thanh Hoa is recently receiving large investments with rapid growth in the coastal areas in the near future. Both case study areas experience cyclones, floods and other hydro-meteorological hazards. Salinization of groundwater and surface water resources, as well as soils, due to sea level rise, huge storms and overexploitation are serious problems for both study areas.

References

COM (Commission of the European Communities) (2009) White paper—adapting to climate change: towards a European framework for action. COM, Brussels. http://eurlex.europa.eu/LexUriServ/LexUriServ.do?uri=CELEX:52009DC0147:en:NOT. Accessed 22 Aug 2012

COM (Commission of the European Communities) (2013) An EU strategy on adaptation to climate change. COM, Brussels. http://www.eea.europa.eu/publications/adaptation-in-europe. Accessed 13 May 2013

IPCC (1994) IPCC technical guidelines for assessing climate change impacts and adaptations. Cambridge University Press, Cambridge

IPCC (2001) Climate change 2001: impacts, adaptation, and vulnerability. Contribution of working group II to the third assessment report of the intergovernmental panel on climate change, Cambridge University Press, Cambridge

IPCC (2007) Fourth assessment report (AR4): climate change 2007. Contribution of working groups I, II and III to the fourth assessment report of the intergovernmental panel on climate change (IPCC), Cambridge University Press, Cambridge

IPCC (2012) Managing the risks of extreme events and disasters to advance climate change adaptation. A special report of working groups I and II of the intergovernmental panel on climate change, Cambridge University Press, Cambridge

IPCC (2014) Climate change 2014: impacts, adaptation, and vulnerability. Part B: Regional aspects. Contribution of working group II to the fifth assessment report of the intergovernmental panel on climate change, Cambridge University Press, Cambridge

Klein J, Jarva J, Frank-Kamenetsky D, Bogatyrev I (2013) Integrated geological risk mapping: a qualitative methodology applied in St. Petersburg, Russia. Environ Earth Sci 70(4):1629–1645. doi:10.1007/s12665-013-2250-1

National Target Program to Respond to Climate Change (Vietnam) (2008) http://www.theredddesk.org/countries/vietnam/info/plan/national_target_program_to_respond_to_climate_change_vietnam. Accessed 26 Apr 2013

Schmidt-Thomé P, Klein J (eds) (2013) Climate change adaptation in practice—from strategy development to implementation. Wiley Blackwell, London, p 327

UNDP (2004) Reducing disaster risk—a challenge for development. United Nations Development Programme, Bureau for Crisis and Recovery, New York

Vietnam Climate Change Adaptation Strategy (2011) http://www.chinhphu.vn/portal/page/portal/English/strategies/strategiesdetails?categoryId=30&articleId=10051283. Accessed 18 Sept 2013

Chapter 2
Climate Change in Vietnam

Abstract Vietnam is considered as one of the countries to be severely affected by climate change and thus response to climate change is of crucial importance to Vietnam, particularly in coastal areas. Rises in average temperatures have been observed over the last decades, as well as substantial changes to precipitation patters. The average temperatures have been rising and the total precipitation has increased, especially during the rainy seasons, which is important for flood water management. In northern Vietnam the precipitation during the dry seasons has decreased, which poses important challenges to water management.

Climate change is one of the most significant challenges facing human being today. Climate change has already affected agricultural production and socio-economic structures and will extensively and intensively alter the development process and security issues including food, water, energy and social safety as well as political, cultural, economic, diplomatic and commercial security (MONRE 2008, 2010, 2012 a, b, 2013).

Vietnam is considered as one of the countries to be severely affected by climate change (IPCC 2001), and thus response to climate change is of crucial importance to Vietnam, particularly in coastal areas. Therefore, the development of suitable adaptation solutions for Vietnamese coastal provinces is extremely essential.

2.1 Recent Climate Change Observations in Vietnam

Vietnam's temperature and precipitation trends have been greatly different among regions during the last 50 years. Annual average temperature has increased by 0.5 °C nationwide and annual precipitation has decreased in the North and increased in the South (see Figs. 2.1 and 2.2) (MONRE 2012a, b, 2013).

© The Author(s) 2015
P. Schmidt-Thomé et al., *Climate Change Adaptation Measures in Vietnam*,
SpringerBriefs in Earth Sciences, DOI 10.1007/978-3-319-12346-2_2

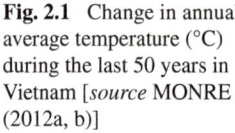

Fig. 2.1 Change in annual average temperature (°C) during the last 50 years in Vietnam [*source* MONRE (2012a, b)]

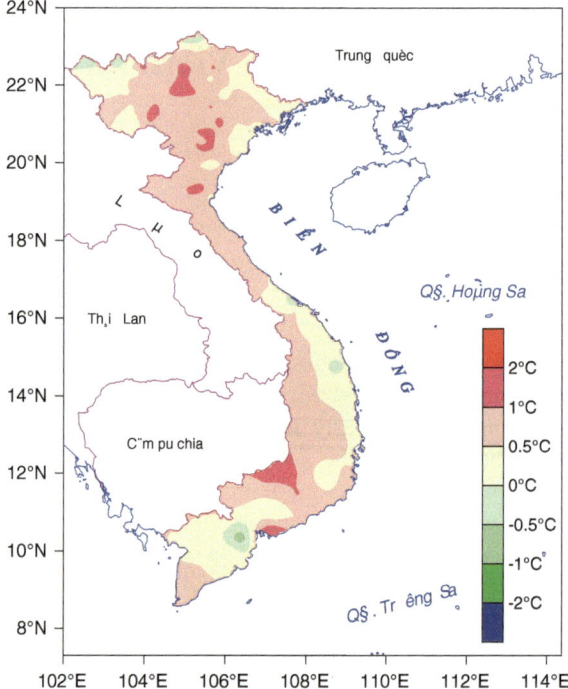

Temperature has increased in almost all regions nationwide, however smaller areas in Central and Southern coastal regions such as Thua Thien–Hue, Quang Ngai, and Tien Giang have seen a decrease in temperature. It is worth noting that precipitation in these areas has increased in both dry and wet seasons.

Changes in maximum temperature in Vietnam varied in the range from −3 to 3 °C. Changes in minimum temperatures mostly varied in the range from −5 to 5 °C. Both maximum and minimum temperatures have tended to increase, with minimum temperatures increasing faster than maximum temperatures, reflecting the trend of global climate warming.

Precipitation during dry season (November–April) has increased slightly or is almost unchanged in the northern regions and increased dramatically in the southern regions during the last 50 years. Precipitation during the rainy season (May–October) has decreased by 5 to over 10 % in most of Vietnam's northern area and increased by 5–20 % in the southern regions. The pattern of change in annual precipitation is similar to the precipitation during the rainy season, i.e. increasing in the southern climate regions and decreasing in the northern climate regions. Annual precipitation in the South Central Region has increased most dramatically compared with other regions in the country during the last 50 years, even by 20 % in some places.

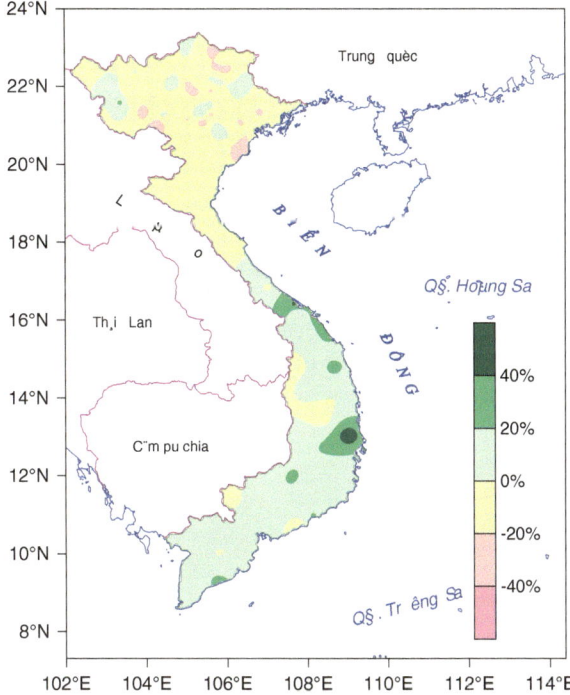

Fig. 2.2 Change in precipitation (%) during the last 50 years in Vietnam [*source* MONRE (2012a, b)]

Table 2.1 Increases in temperature and changes in precipitation during the last 50 years in the climate regions of Vietnam

Climate regions	Temperature (°C)			Precipitation (%)		
	January	July	Year	Nov–Apr period	May–Oct period	Year
Northwest	1.4	0.5	0.5	6	−6	−2
Northeast	1.5	0.3	0.6	0	−9	−7
Red river delta	1.4	0.5	0.6	0	−13	−11
North central coast	1.3	0.5	0.5	4	−5	−3
South central coast	0.6	0.5	0.3	20	20	20
Central highlands	0.9	0.4	0.6	19	9	11
South region (southeast and Mekong delta)	0.8	0.4	0.6	27	6	9

Source MONRE (2012a, b)

Maximum daily precipitation has increased in almost all regions, particularly during the recent years. The number of days with heavy rain seems to have increased similarly, with significant changes in the central regions (MONRE 2012a, b, 2013) (Table 2.1).

In terms of tropical cyclones there are annually on average about 12 tropical storms on the East Sea (Vietnamese name for the South China Sea), of which about 45 % form in the East Sea and 55 % stem from the Pacific Ocean. There are about seven storms affecting Vietnam annually, five of which hit or directly affect the country's mainland. The areas where storms or tropical depressions occur most frequently are in the middle of the Northern East Sea. In the central coastal area from 16°N to 18°N and in the northern coastal area from 20°N northwards, the storms or tropical depressions occur most frequently along the coastal region, with a storm or tropical depression landfall every 2 year in average (see Fig. 2.3).

The number of tropical cyclones occurring in the East Sea seems to have increased slightly, while the number of cyclones affecting or hitting Vietnam's mainland does not have an obvious changing pattern. Over the past decades,

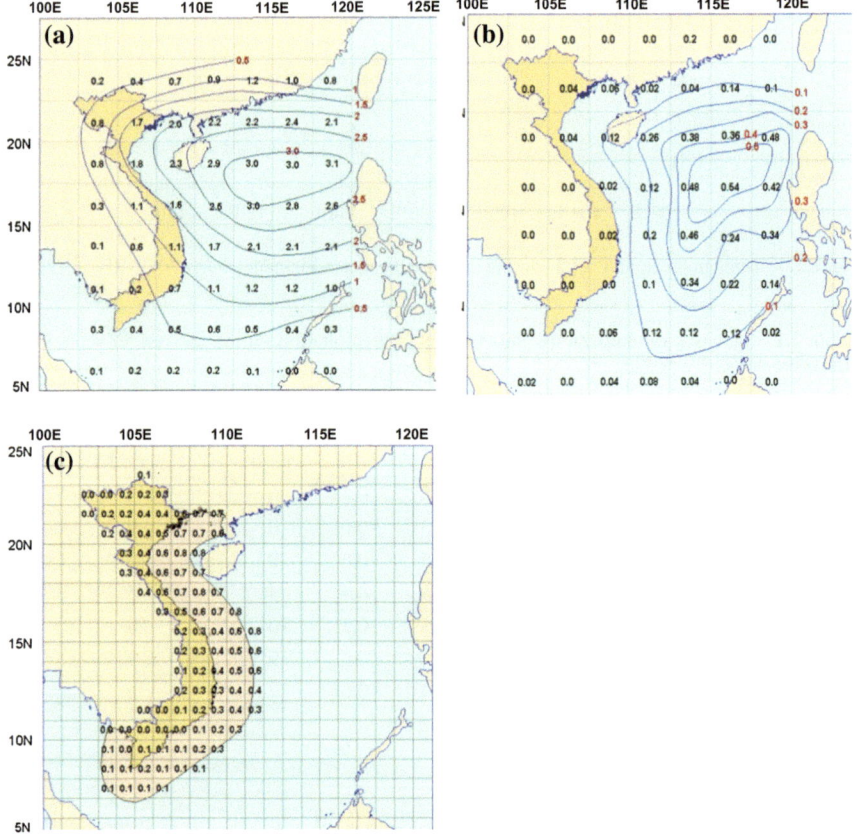

Fig. 2.3 Map of recurrence of tropical cyclones operating (**a**) and forming in the east sea (**b**), and hitting Vietnam's mainland (**c**) [*source* MONRE (2012a, b)]

tropical storms approaching Vietnam seem to move southwards, the number of very strong storms seems to increase and the storm season tends to last longer. Overall, storm impacts on the country seem to increase (MONRE 2012a, b) (Fig. 2.4).

Drought, including monthly and seasonal drought, seems to increase in differing patterns among the climate regions. Hot and sunny weather has shown signs of increasing dramatically in various regions in the country, particularly in the North Central and in southern regions (MONRE 2012a, b).

The water level monitored at Vietnam coastal gauges has shown that the pattern of changes in annual average sea level is different over the years (starting 1960) (Fig. 2.5). Almost all the stations have shown an increasing trend, however, some stations do not show this trend. Based on data collected from the monitoring stations the mean sea level rise along the Vietnamese coastal area is about 2.8 mm/year during the period 1993–2008 (MONRE 2009).

Sea level data derived from satellite images from 1993 to 2010 have shown that the change in sea level in the whole East Sea is 4.7 mm/year, and it has increased quicker in the east than in the west of the East Sea. Along Vietnam's coastal zone, the sea level in the mid-central coastal region and southwestern region seems to have increased more, with average sea level rise of the whole coastal zone of about 2.9 mm/year approximated from satellite images (MONRE 2009).

The sea level pattern for the coastal zone is almost the same using actual measured data at coastal gauges and data derived from satellite images. This has been verified when comparing actual measured data at coastal gauges and satellite data. Comparison results have shown a good correlation of phase and amplitude of average water level variations (MONRE 2012a, b).

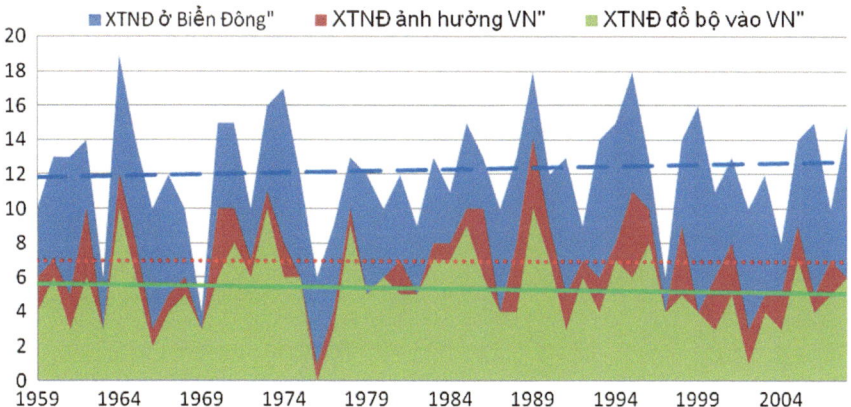

Fig. 2.4 Patterns of the number of tropical cyclones operating (*blue*) and affecting (*red*) in the east sea, and hitting (*green*) Vietnam's mainland during the last 50 years [*source* MONRE (2012a, b)]

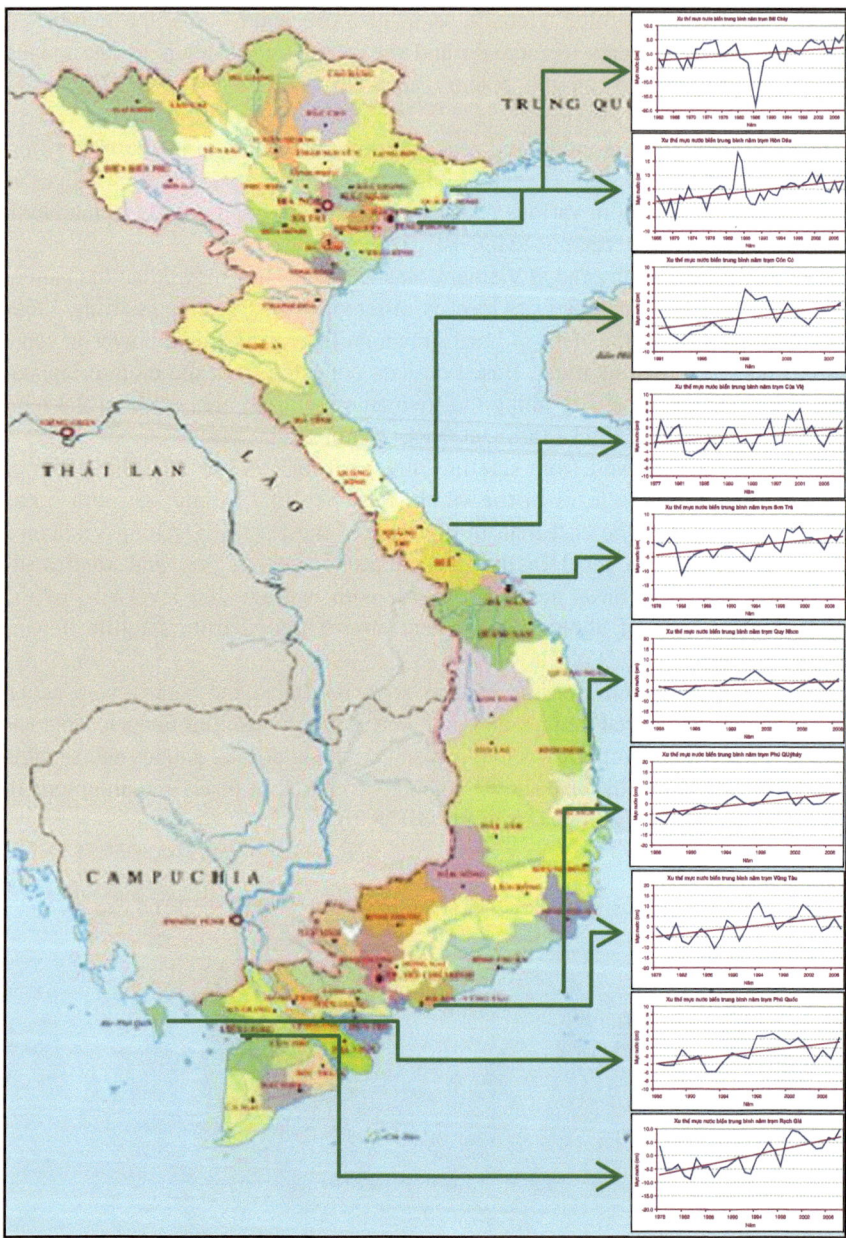

Fig. 2.5 Change in sea level based on data from measuring stations (1960–2010) [*source* MONRE (2009, 2012a, b)]

2.2 Climate Change Scenarios for Vietnam

Climate change and sea level rise scenarios developed for Vietnam are based on different greenhouse gas emission scenarios of IPCC's 4th report (IPCC 2007), namely a low scenario (B1), medium scenarios (B2, A1B), and scenarios of the high anthropogenic greenhouse gas emission (A2, A1FI).

Climate change scenarios for temperature and precipitation are developed for seven climate regions in Vietnam: North West, North East, Northern Delta, North Central Region, South Central Region, Central Highlands, and Southern Region.

Temperatures at the end of 21st century in Vietnam (MONRE 2012a, b)

- Low emission scenario (B1): annual average temperature increases by 1.6–2.2 °C.
- Medium emission scenario (B2): annual average temperature increases by 2–3 °C.
- High emission scenario (A2, A1F1): annual average temperature increases by 2.5–3.7 °C.

Precipitation at the end of 21st century in Vietnam (MONRE 2012a, b)

- Low emission scenario (B1): annual precipitation increases 2–6 %.
- Medium emission scenario (B2): annual precipitation increases 2–7 %.
- High emission scenario (A2, A1F1): annual precipitation increases 2–10 %.

Sea level rise at the end of 21st century in Vietnam (MONRE 2012a, b)

- Low emission scenario (B1): average sea level may increase by 49–64 cm.
- Medium emission scenario (B2): average sea level may increase by 57–73 cm.
- High emission scenario (A2, A1FI): average sea level may increase by 78–95 cm

Climate extremes in Vietnam
By the end of the 21st century, the number of days with maximum temperature of over 350 °C increases from 15 to 30 days in almost all regions in the country based on the medium emission scenario A1B (MONRE 2012a, b).

In the future, the general trend is that the maximum daily precipitation increases in the North West, North East, Red River Delta and North Central Regions, and decrease in the South Central Coastal, Central Highlands, and the South Regions (MONRE 2012a, b).

If the sea level rises by one meter, about 39 % of the Mekong river delta area (MRD), over 10 % of the Red River delta, over 10 % of the Quang Ninh province, 20 % of the Ho Chi Minh City area (HCMC) are flooded, which constitute 6.3 of the total land area. According to this scenario 35 % of the MRD population and 7 % of HCMC population will be affected (MONRE 2012a, b, 2013).

2.3 Climate Change Impacts on Water Resources

Climate change has effects on the average river runoff and annual distribution of peak flows (both high and low). The impacts of climate change on these flow patterns in future periods are simulated by the rainfall-runoff modelling under different climate change scenarios.

Impacts of climate change on annual flows vary between regions and river systems across Vietnam. In the medium CC scenario (B2), the annual flow of rivers in the Red River Delta and the northern part of the North Central region tend to increase by less than 2 % in 2040–2059 and by from 2 to 4 % in the period 2080–2099. In contrast, the projected annual flow of rivers from the southern part of the North Central region to the northern part of the South Central Region and the Southeastern region (Dong Nai river system) tend to decrease at different levels, by 2 % in Thu Bon river and Ngan Sau river, and more significantly by 4–7 % in the period 2040–2059 in the Dong Nai river and the Be river basins, and up to 7–9 % in the period 2080–2099.

Flood events of most rivers tend to increase by 2–4 % in the period 2040–2059 and by 5–7 % in the period 2080–2099, but to varying degrees. In the Thu Bon and Ngan Sau Rivers, flood flow tends to increase by less than 2 % in the period 2040–2059 and less than 3 % in the period 2080–2099.

After 2020, groundwater levels may decrease considerably due to both overexploitation and decreased groundwater recharge during the dry season. In the south, if the flow of the rivers decreases by 15–20 % in the dry season, the corresponding groundwater level may decrease by 11 m as compared to the current level (MONRE 2013).

References

IPCC (2001) Climate change 2001: impacts, adaptation, and vulnerability. Contribution of working group II to the third assessment report of the intergovernmental panel on climate change, Cambridge University Press, Cambridge

IPCC (2007) Fourth assessment report (AR4): climate change 2007. Contribution of working groups I, II and III to the fourth assessment report of the intergovernmental panel on climate change (IPCC), Cambridge University Press, Cambridge

MONRE (Bộ Tài Nguyên Môi Trường) (2008) Chương trình Mục tiêu ứng phó với BĐKH, Hà Nội, 7/2008 (Ministry of Natural Resources and Environment (2008) Target programme to respond to climate change, Hanoi, 7/2008)

MONRE (Bộ Tài Nguyên Môi Trường) (2009) Kịch bản biến đổi khí hậu, nước biển dâng cho Việt Nam. Hà Nội (Ministry of Natural Resources and Environment (2009) Climate change scenario, sea level rise for Vietnam, Hanoi)

MONRE (Bộ Tài Nguyên Môi Trường) (2010) Sổ tay hướng dẫn Công cụ phân tích biến đổi khí hậu—Chương trình hợp tác chính phủ Việt Nam và Đức GTZ và IFAD (Ministry of Natural resources and the Environment (2010) Handbook on analysis tools to guide climate change—partnership program Vietnam government and the German GTZ and IFAD)

MONRE (Bộ Tài Nguyên Môi Trường) (2012a) Kịch bản Biến đổi khí hậu, nước biển dâng cho Việt Nam. Hà Nội (Ministry of Natural Resources and Environment (2012) Climate change scenarios and sea level rise for Vietnam, Hanoi)

MONRE (Bộ Tài Nguyên Môi Trường) (2012b) 2 Chiến lược quốc gia về BĐKH. Hà Nội (Minister of Natural Resources and Environment (2012) National strategy on climate change, Hanoi)

MONRE (Bộ Tài Nguyên Môi Trường) (2013) Tài liệu hướng dẫn đánh giá tác động của biến đổi khí hậu và các giải pháp thích ứng, NXB Tài nguyên—Môi trường và Bản đồ Việt nam (Ministry of Natural Resources and Environment (2013) Documentation to assess the impact of climate change and adaptation measures, Hanoi)

Chapter 3
Impacts of Climate Change on the Thanh Hoa Province

Abstract The most important climate change impacts in the Thanh Hoa province are increasing temperatures and precipitation during the rainy season. Dry seasons on the other hand have been decreasing precipitation which actually leads to water shortages in some areas. The rising sea level causes salinization of coastal aquifers. Along with this goes a strong socio-economic development of the entire province, leading to a higher water demand as well as risks to recharge areas. The detailed assessment of aquifers as well as water balance calculations has been started as a basis to build scenarios for future water supply. These scenarios take both climate change models and the impacts of socio-economic development into account and support the development and implementation of climate change adaptation strategies in close cooperation with local stakeholders.

Thanh Hoa is a province in the North Central Region, with the geographical coordinates of 19°23′–20°30′ northern latitude, 104°23′–106°30′ eastern longitude. The total area is 1,112,032.83 ha, which accounts for 3.37 % of total natural area of the country.

The province borders three provinces of Ninh Binh, Hoa Binh, and Son La to the north, Nghe An province to the south, Hua Phan province (Lao PDR) to the west with a border line of 192 km and the East Sea to the east with a coastline of 102 km.

The maximum extent from west to east is 110 km and from north to south 100 km. The province's economic and political center is the city of Thanh Hoa, which is 153 km to the south of Hanoi. Of the 3.4 million people living in the Thanh Hoa province about 200,000 reside in the city of Thanh Hoa (Cục thống kê tỉnh Thanh Hóa 2010) (Fig. 3.1).

© The Author(s) 2015
P. Schmidt-Thomé et al., *Climate Change Adaptation Measures in Vietnam*,
SpringerBriefs in Earth Sciences, DOI 10.1007/978-3-319-12346-2_3

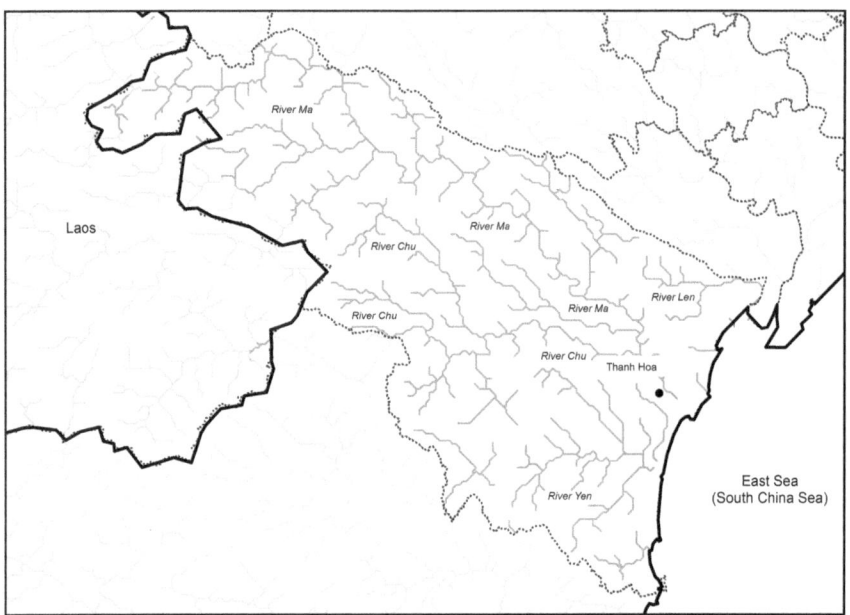

Fig. 3.1 Map of Thanh Hoa province

3.1 State of the Environment in Thanh Hoa Province

Ambient air quality in Thanh Hoa province is quite good. All the quality criteria such as suspended dust, noise, SO_2, NO_2 and CO are within permitted limits in accordance with Vietnamese Standards (QCVN 05: 2013/BTNMT).

The air in industrial zones (IZ) have shown signs of pollution with NO_2, SO_2 and suspended dust. However, the signs of pollution in these areas seem to have decreased recently. Air in industrial clusters and trade villages have been polluted with SO_2 (exceeding 1–2 times Vietnamese air quality standards) and suspended dust (exceeding 1.08–12.8 times Vietnamese standards). Surface water in almost all the main river systems in the province has been slightly polluted, particularly by COD, BOD, TSS, NH_3, oil products and coliform bacteria. Specifically, the COD values at all the downstream locations exceed the Vietnamese surface water quality standards (QCVN 08: 2008/BTNMT) and are much higher than those at upstream locations.

In the Ma river system, COD values at all observation stations exceed the Vietnamese quality standards (between factors 1.3 and 7.2). BOD values at measurement points are mostly lower than the limit values in the Vietnamese quality standards. Total suspended solids (TSS) exceed the Vietnamese quality standards at all observation points. In addition, pollution due to NH_3 and oil products seem to increase and show signs of exceeding the Vietnamese quality standards.

Coastal sea water in Thanh Hoa province has quite high total suspended solids (TSS) values. In the rainy season, TSS values at coastal survey stations (up to 8 m depth) exceed the limits of Vietnamese quality standards. Manganese and Fe contents are 1.2–2.6 times higher than the permitted limits for sea water for aquaculture and concentrations of oil products are higher than the permitted limit values for coastal sea water for aquaculture and beaches (Trung 2010).

3.2 Hydrogeology

The Holocene aquifer (qh)
The Holocene aquifer (qh) comprises the two water bearing layers qh2 and qh1. The qh1 water bearing layer consists of fluvial and/or marine sediments. The layer is between 5 and 22 m thick and its water retention capacity is low.

The qh2 water bearing layer is an important groundwater resource, particularly in coastal areas, where almost all surface waters and deep water aquifers are saline. This layer consists of sediments from a variety of sources: fluvial, marine and eolic and is between 2 and 25 m thick. The lithologic composition is mainly silt and fine-grained sand. The Holocene aquifer is recharged by precipitation and by surface water infiltration.

Discharge rates of the test wells range from 0.06 to 2.63 L/s, with an average value of 1.1 L/s. The hydraulic conductivity (K) ranges between 1.29 and 11.78 m/day, with an average value of 4.56 m/day (which is a typical value for fine sand). The respective storativity (S) is between 0.0002 and 0.28, averaging 0.07.

Results of the national water resource monitoring network in 2011 show the following hydrogeological characteristics in the Thanh Hoa Province.:

- The depth of the water level in the qh aquifer varies from 0.5 to 3.8 m. The water level varies depending on the seasons, with lower water levels during the dry season. The annual variation is from 0.9 to 3 m.
- Total dissolved solids (TDS) levels in the monitoring wells are always lower than the permitted groundwater quality limit value, 1,500 mg/l. The highest value is 1.34 mg/l at QT13-TH monitoring well (Nga Son).
- Concentrations of inorganic substances were determined at nine monitoring wells. Analysis results have shown that almost all parameters have lower contents than the permitted groundwater quality standards, except for Mn and As.
- Six monitoring wells showed a Mn content higher than the permitted groundwater limit value, 0.5 mg/l. The highest Mn content of 1.87 mg/l is found at QT12-TH monitoring well (Quang Chinh, Quang Xuong).
- In one monitoring well in the Thieu Hoa district a higher As content, (0.07 mg/l) than the permitted groundwater limit value, 0.05 mg/l is found.
- Ammonia parameters are measured from 11 monitoring wells. According to analysis results, the contents in all monitoring wells are higher than the

permitted groundwater limit value (0.1 mg/l calculated as nitrogen). The highest content is 17.10 mg/l in QT12-TH monitoring well (Hoang Hoa).

The Pleistocene aquifer (qp)

The pleistocene aquifer is mainly formed of Hanoi formation sediments (Q_1^{2-3}hn). The aquifer is not exposed on the surface and appears in almost all deep boreholes in the region.

The lithologic composition comprises pebbles and sand with thicknesses varying from 9.3 (LK3HR) to 51.8 m (LK9HR). The depth of the qp aquifer reaches from 7.3 to 88 m below ground with an average depth of 30 m.

Water levels in the Holocene and Pleistocene aquifers are similar. The Pleistocene aquifer is recharged by precipitation, interflow and groundwater infiltration from the river systems.

The water retention capacity of the qp aquifer is good with discharge rates between 9 and 61 L/s (LK18 and LK6, respectively). The drawdown varies from 1.09 to 11.6 m (LK10 and LK22) and the specific capacity is between 1.27 and 21.87 (L/s)/m (LK4 and LK10). The transmissivity varies from 657 m^2/day (LK2HR) to 2,100 m^2/day (LK6HR).

Results of the national water resource monitoring network in 2011 show the following water quality characteristics in Thanh Hoa province:

- The TDS value is monitored in 12 monitoring wells. Analysis has shown that TDS values in two monitoring wells are higher than the permitted limit value, 1,500 mg/l. The highest value is 14.26 mg/l in the QT9a-TH monitoring well (Sam Son town).
- Inorganic substances are determined at 12 monitoring wells. Analysis results have shown that almost all parameters have lower contents than the permitted groundwater quality standards, except for Mn and As. The Mn content is higher than the permitted limit value (0.5 mg/l) in eight facilities. The highest Mn content of 1.93 mg/l is found in the QT9a-TH monitoring well (Truong Son precinct, Sam Son town). The As content in four monitoring wells is higher than the permitted limit value, 0.05 mg/l. The highest As content of 0.22 mg/l is found in the QT1a-TH monitoring well (Yen Dinh).
- Ammonia parameters are monitored at 10 monitoring wells. According to analyses, the contents in all wells are higher than the permitted limit value (0.1 mg/l, calculated as nitrogen). The highest content is 7.95 mg/l in the QT5a-TH montoring well (Tho Xuan).

Fissured aquifers

The fissured aquifers (including karst) are distributed in the plain margin, hills and scattered mountains and in deep layers. In these aquifers, water flows in the tectonic and weathered fissures and karst caves. The studies are still preliminary but generally the water retention capacity is low in fissured aquifers.

3.3 Groundwater Reserves

Thanh Hoa plain

The Holocene aquifer (qh)
The estimated natural recharge of the qh aquifer in the Thanh Hoa plain is 503.918 m^3/day for the area of 2,237 km^2. The average recharge is 1 7 (L/s)/km^2, leading to a total recharge (Q) of 328,570 m^3/day (Q = 2,237 × 10^6 × 1.7 × 86,400/10^6 = 328,570 m^3/day).

The potential reserve of the qh aquifer is 391,475,000 m^3 with an average thickness of 5 m and an average specific yield (S$_y$) of 0.035.

The Pleistocene aquifer (qp)
The estimated natural dynamic reserve of the qp aquifer in the Thanh Hoa plain is calculated for the area of 2,237 km^2, where the average recharge is 1.6 (L/s)/km^2, leading to a total recharge (Q) of 309,242 m^3/day (Q = 2,237 × 10^6 × 1.6 × 86,400/10^6 = 309,242 m^3/day).

The static reserve of the qp aquifer in Thanh Hoa plain is 823,216,000 m^3/day for the area of 2,237 km^2 with an average aquifer thickness of 23 m and an average specific yield of 0.016.

Thanh Hoa coastal area (model area)

The Holocene aquifer (qh)
The natural dynamic reserve of the qh aquifer in the Thanh Hoa coastal area is 177,871 m^3/day for the area of 1,211 km^2, with an average supply of 1.7 (L/s)/km^2.

The potential reserve of the qh aquifer is of 211,925,000 m^3 with an average thickness of 5 m and an average specific yield of 0.035.

The Pleistocene aquifer (qp)
The natural dynamic reserve of the qp aquifer in Thanh Hoa coastal area is 167,408 m^3/day for the area of 1,211 km^2. An average recharge is 1.6 (L/s)/km^2.

The static reserve of the qp aquifer in the Thanh Hoa coastal area is 445,648,000 m^3/day for the area of 1,211 km^2. The average aquifer thickness is 23 m and the average specifics yield is 0.016 (Fig. 3.2).

Socio-economy and development objectives
The Thanh Hoa province has one city (grade II city), two district-level towns (grade IV cities), 24 districts (13 lowland districts, 11 mountainous districts), 30 commune-level towns (24 country towns, 6 other industrial-service-commercial towns) and 20 precincts and 584 communes. In 2010, Thanh Hoa's population was 3,406,805 people, with a natural growth rate of 0.7 %. The population density was of 306 people/km^2, and about 10 % of the population live in urban areas.

At present, the size of the province's economy is not yet proportional to the size, and development potential of the province as people's income is comparatively low. One challenging factor is the infrastructure as settlements are spread out and difficult to reach, particularly in remote and mountainous areas. The province's total GDP reaches VND 51,000 billion (ca 1.75 billion €), accounting for 3 % of the nation's GDP.

HYDROGEOLOGY MAP
COASTAL AREA IN THANH HOA

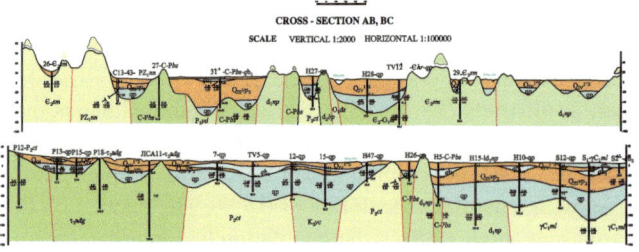

Fig. 3.2 Hydrogeologic map of the coastal area of the Thanh Hoa province

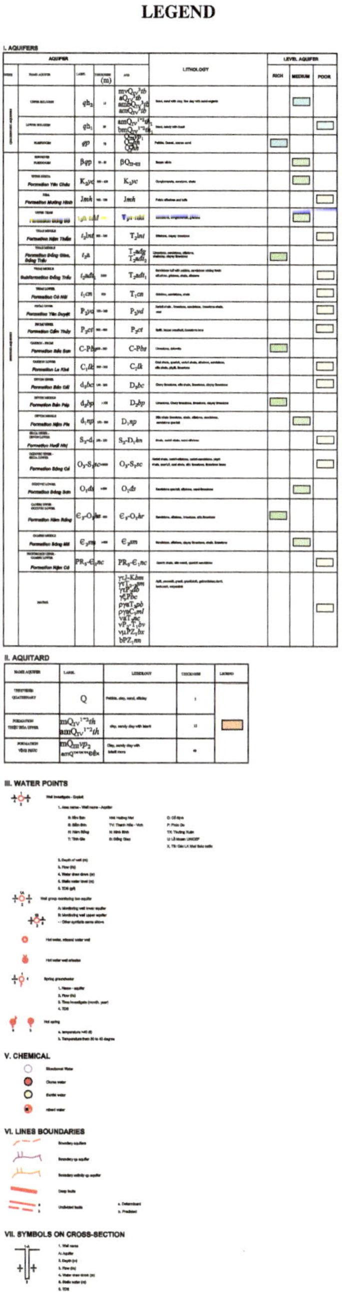

LEGEND

Fig. 3.2   (continued)

Together with the growth rate, Thanh Hoa's economic structure has gradually shifted towards a higher proportion of industry and construction in the total GDP and lower proportion of agriculture, forestry, and fishery.

Thanh Hoa province's economic structure is as following: Industry and construction account for 41.5 %, tourism and services 34.4 % and agriculture, forestry, and fishery 24.1 % of the total GDP (Cục thống kê tỉnh Thanh Hóa 2010).

Industry and construction
Total industrial production in real price reaches VND 32,000 billion (1.1 billion €), of which the state economic sector contributes 26 %. The non-state economic sector contributes 54.6 %, and foreign direct investment reaches 19.4 %.

Tourism and services
Total commodity retail sales and service turnover in real price reaches over VND 23,000 billion (790 million €).

In terms of tourism activities, there are over 711 accommodation establishments, including 80 hotels and 631 hostels. Total number of rooms is over 11,737, the number of beds is of about 22.4 thousand. The number of annual accommodations in hotels is about 2.1 million of which 0.05 % are foreign visitors (Cục thống kê tỉnh Thanh Hóa 2010).

Agriculture, forestry, and fishery
The province's agricultural production in real price reaches over VND 17,774.2 billion. Cultivation has a share of 70.6 %, breeding contributes 26.5 %, and the service sector 2.9 % of the total agricultural production.

Socio-economic development objectives until 2020
For the Thanh Hoa province, the average annual growth rate for the period 2011–2015 is approximated to be 17–18 % and over 19 % for the period 2016–2020. As of 2015, the GDP per capita should reach the national average level, and it is expected to exceed it after 2015. The economic structure is shifting towards industrialization. The economic structure for 2015 is estimated as following: agriculture 15.5 %, industry and construction 47.6 %, service sector 36.8 %. The projections for 2020 are 10.1, 51.9 and 38 %, respectively. Export turnover is envisaged to reach 620 million € in 2015 and over 1.4 billion € in 2020 with an export growth rate of 19–20 % per year. Budget revenue accounts for 6–7 % of GDP in 2015 and over 7 % in 2020. The natural population growth rates are under 0.65 % in 2015 and about 0.5 % in 2020.

(*Source: Decision No. 144/ 2009/QD-TTg dated 28/9/2009 on the approval of Socio-Economic Development Plan of Thanh Hoa province until 2020*).

3.4 Impacts of Socio-economic Development and Climate Change on the Environment

The socio-economic development impacts on the environment include mainly urbanization and industrialization, as well as agricultural activities, tourism development and service development.

Urbanization process

The population of the province is growing in both total amount and density, especially in the coastal areas. This leads to increases in the demand and the exploitation of natural resources, and leads to increases in domestic sewage and solid wastes.

Most of domestic sewage is only treated primarily in septic tanks and then discharged into the culvert system or canals, ponds, or lakes. However, the septic tanks are not considered as very effective due to low technical standards, and thus the levels of polluted substances in wastewater are very high. Most domestic sewage from residential areas is polluted with organic matter (COD, BOD), suspended solids and Coliform bacteria. The five urban areas of the province have a total estimated volume of sewage of 47,190 m³/day (domestic sewage accounting for about 31,800 m³/day, industrial production 12,390 m³/day and other services 3,000 m³/day).

The urban development process also causes an increase in solid wastes discharged to the environment. At present, only about 70 % of the city garbage is collected with capacity of about 120 tons/day (43,200 tons/year).

Industrialization process

Industrial development is accompanied with the increase in consumption of raw materials, natural resources (especially water), land consumption and soil sealing, as well as air pollution.

With rapid socio-economic development, particularly in the industrial sector, the Thanh Hoa province is facing a variety of challenges in wastewater treatment and industrial solid waste management, both of which might lead to substantial environmental pollution. At present, almost all the industrial zones, industrial clusters and trade villages in the province poses, if any at all, only preliminary wastewater treatment systems.

Currently, the total volume of industrial solid waste is of about 48,000 tons/year. Trade villages producing marble stone products discharge large volumes of marble powder. Since there has been no solution for recycling or reusing this waste, it must be buried. At present, Dong Son district has reserved 2 ha in Dong Hung commune for marble powder landfill. Industrial solid wastes are mostly being treated by the respective waste owner.

Industrial wastewater and solid waste pollution is a threat for the environment without centralized wastewater and garbage treatment system for industrial activities in this area (Trung 2010).

3.5 Impacts of Climate Change

Temperature

According to monitoring data of 1980–2009, the average temperature in Thanh Hoa in this period was of about 23.8 °C. Measurements revealed that the average temperature tends to increase at a speed corresponding to 0.011 °C/year.

In this report, we apply two climate change scenarios for further analysis, the medium emission scenario B2 (according to the recommendation of MONRE) and high emission scenario A1FI.

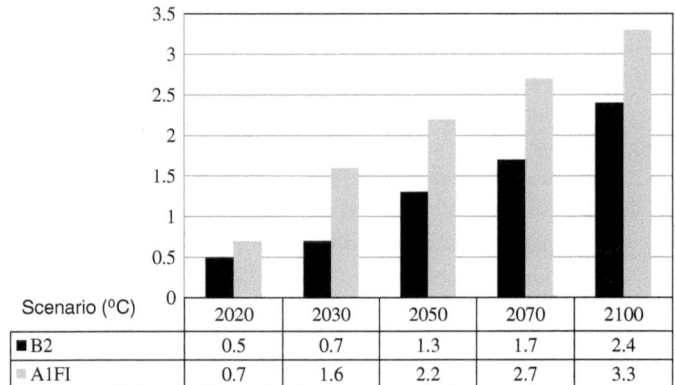

Scenario (°C)	2020	2030	2050	2070	2100
■ B2	0.5	0.7	1.3	1.7	2.4
▨ A1FI	0.7	1.6	2.2	2.7	3.3

Fig. 3.3 Changes of mean temperatures (°C) compared to the period 1980–2009 under climate change scenarios B2 and A1FI in the Thanh Hoa province from 2020 to 2100

Results from calculations of the climate change scenarios B2 and A1FI show that the temperature tends to gradually increase to 1.3–2.2 °C by 2050, and further increase to 2.4–3.3 °C by 2100. In terms of distribution of future average temperatures in Thanh Hoa, there is a difference between the province's regions, i.e. changes in temperature are increasing from coastal regions to the mountainous regions of Thanh Hoa province (Nguyễn et al. 2010) (Figs. 3.3 and 3.4).

Precipitation

According to precipitation data from 1980 to 2009, the average precipitation in coastal areas has been decreasing by −13.93 mm/year. In central Thanh Hoa, in Lang Chanh, the decrease was of −11.13 mm/year, and −0.7 mm/year in Cam Thuy. On the other hand, in the Hoi Xuan station (located ca 100 km inland from the city of Thanh Hoa) precipitation increased by +1.4 mm/year.

Projected from climate change scenarios B2 and A1FI, precipitation in Thanh Hoa tends to increase by 2050 with 3.1 and 3.9 %, respectively, and by 2100 with 5.9 and 6.8 %, respectively compared to the baseline period 1980–2009 (Fig. 3.5).

In terms of precipitation changes of the southern districts, the precipitation increase in this region is higher than in other regions of the province, i.e. from 1.9 to 2.3 % for scenario B2 and 2.3–3 % for scenario A1FI by 2050. The precipitation increase reduces gradually towards the North and North West regions of the province, where the change is modeled to vary from 1.3 to 1.7 % for the scenario B2 and 1.6–2 % for the scenario A1FI by 2050 (Nguyễn et al. 2010) (Fig. 3.6).

Impacts of sea level rise

According to the sea level data measured at Ngoc Tra station between 1962 and 2009, the annual average sea level rise rate in the Thanh Hoa province is 1.3 mm/year. According to the medium emission scenario B2 sea level is projected to rise 24 cm by 2050, and 65 cm by 2100 in Thanh Hoa. Based on the projections of the high emission scenario A1FI, the sea level rises 27 cm by 2050 and 86 cm by 2100 (Fig. 3.7).

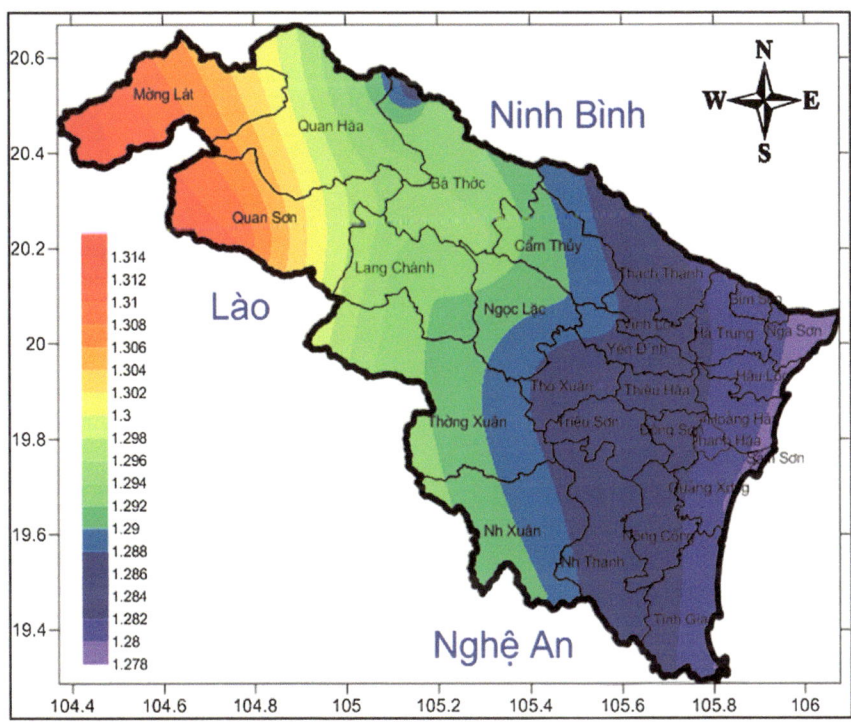

Fig. 3.4 Changes in mean temperature (°C) compared to period 1980–2009 in the Thanh Hoa province in 2050 projected for the scenario B2

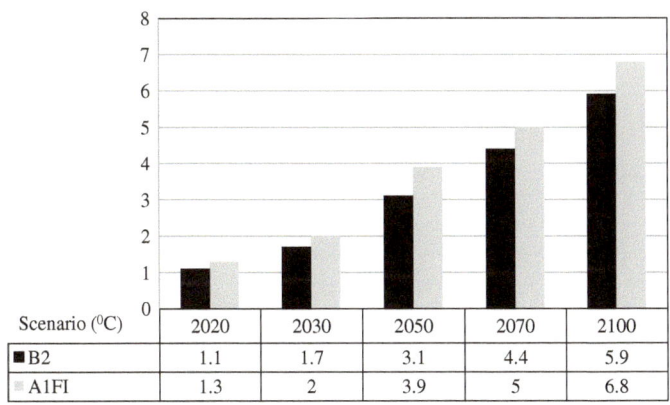

Scenario (°C)	2020	2030	2050	2070	2100
■ B2	1.1	1.7	3.1	4.4	5.9
A1FI	1.3	2	3.9	5	6.8

Fig. 3.5 Change in mean precipitation (%) compared to period 1980–2009 under climate change scenarios B2 and A1FI in the Thanh Hoa province from 2020 to 2100

Results from calculations of sea level rise impacts on land areas show that in the Thanh Hoa province 3.1 % of the area might be under water when sea level rises by 1 m (Source: Climate change scenario and SLR, MONRE 2012).

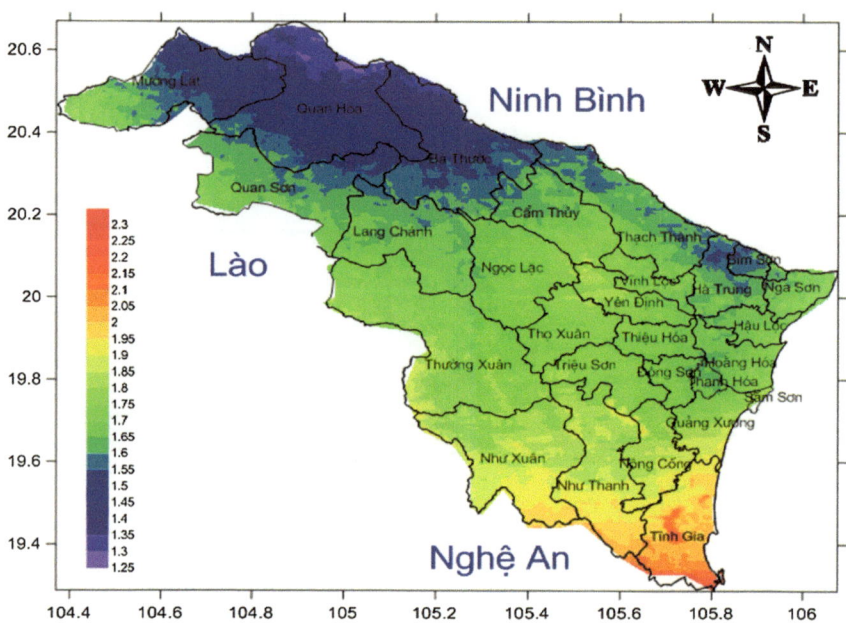

Fig. 3.6 Change in mean precipitation compared to period 1980–2009 in the Thanh Hoa province in 2050 projected for scenario B2

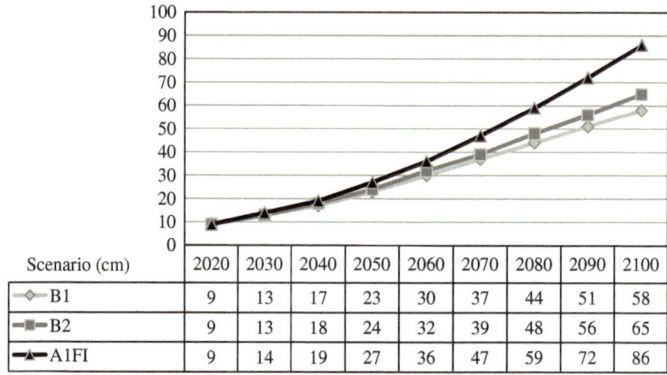

Scenario (cm)	2020	2030	2040	2050	2060	2070	2080	2090	2100
B1	9	13	17	23	30	37	44	51	58
B2	9	13	18	24	32	39	48	56	65
A1FI	9	14	19	27	36	47	59	72	86

Fig. 3.7 Sea level rise scenario (cm) in the Thanh Hoa province projected by emission scenarios B1, B2 and A1FI

Coastal districts have a high flood hazard potential from storm surges. Affected are some densely populated areas (e.g. Sam Son) but most of these areas are currently mainly under agricultural land use (partly up to 50 %) (Nguyễn et al. 2010).

Impacts of climate change on surface water resources

According to the climate change scenarios B2 and A1FI, the annual average precipitation in the Thanh Hoa province tends to increase by 2020, 2050 and 2100, as

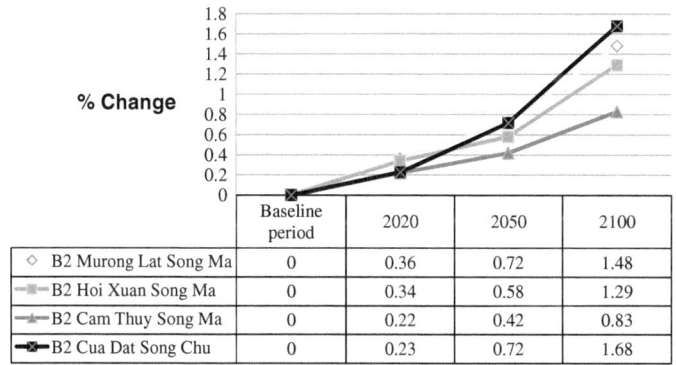

	Baseline period	2020	2050	2100
◇ B2 Murong Lat Song Ma	0	0.36	0.72	1.48
■ B2 Hoi Xuan Song Ma	0	0.34	0.58	1.29
▲ B2 Cam Thuy Song Ma	0	0.22	0.42	0.83
✕ B2 Cua Dat Song Chu	0	0.23	0.72	1.68

Fig. 3.8 Change in annual average flow (%) at monitoring stations of river Ma and river Chu under scenario B2 from 2020 to 2100 compared to baseline period (1980–2009)

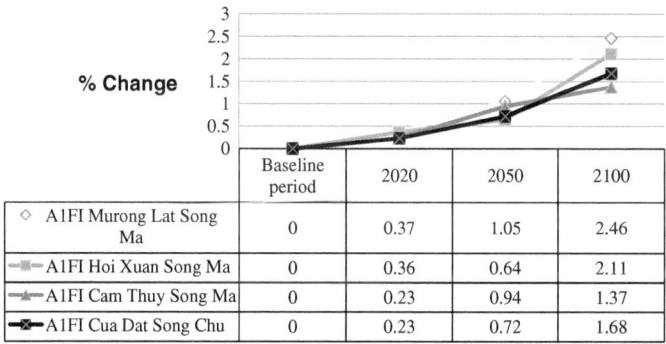

	Baseline period	2020	2050	2100
◇ A1FI Murong Lat Song Ma	0	0.37	1.05	2.46
■ A1FI Hoi Xuan Song Ma	0	0.36	0.64	2.11
▲ A1FI Cam Thuy Song Ma	0	0.23	0.94	1.37
✕ A1FI Cua Dat Song Chu	0	0.23	0.72	1.68

Fig. 3.9 Change in annual average flow (%) at monitoring stations of river Ma and river Chu under scenario A1FI from 2020 to 2100 compared to the baseline period (1980–2009)

compared to the baseline period (1980–1999). Calculations show that the annual average surface water flows follow the increasing tendency in accordance with the trends of precipitation. Specifically, flows at Ma and Chu river gauging stations increase by up to 0.37 % by 2020, 1.05 % by 2050 and up to 2.46 % by 2100 (scenario A1FI) (see Figs. 3.8 and 3.9).

Average flows in flood seasons

The climate change and SLR scenarios of MONRE (2012) all show that the precipitation during rainy seasons tends to increase in the Thanh Hoa province. Average flows in flood seasons in the river systems of the Thanh Hoa province deliver corresponding results of increased runoff. More specifically, flows in stations of Muong Lat, Hoi Xuan, Cam Thuy and Cua Dat have different growths in various periods as illustrated in the Figs. 3.10 and 3.11.

Calculation results show that flows at all gauging stations increase, the average growth rates in both scenarios (B2 and A1FI) are about 0.6 % by 2020, 2 % by 2050 and 4 % by 2100 as compared to the baseline period (1980–1999).

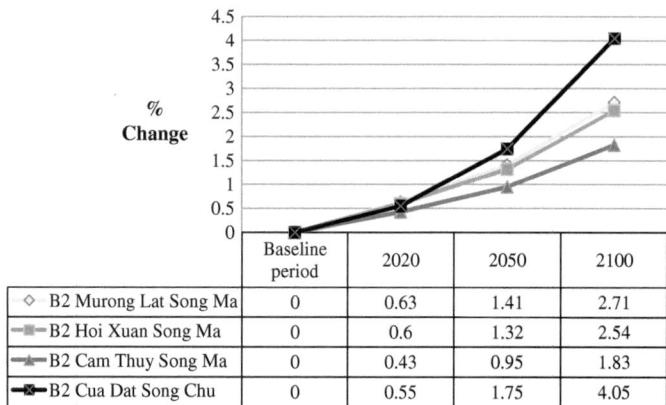

	Baseline period	2020	2050	2100
◇ B2 Murong Lat Song Ma	0	0.63	1.41	2.71
━■━ B2 Hoi Xuan Song Ma	0	0.6	1.32	2.54
━▲━ B2 Cam Thuy Song Ma	0	0.43	0.95	1.83
━✖━ B2 Cua Dat Song Chu	0	0.55	1.75	4.05

Fig. 3.10 Change in average flow in rainy season (%) at monitoring stations of river Ma and river Chu under the scenario B2 from 2020 to 2100 compared to the baseline period (1980–2009)

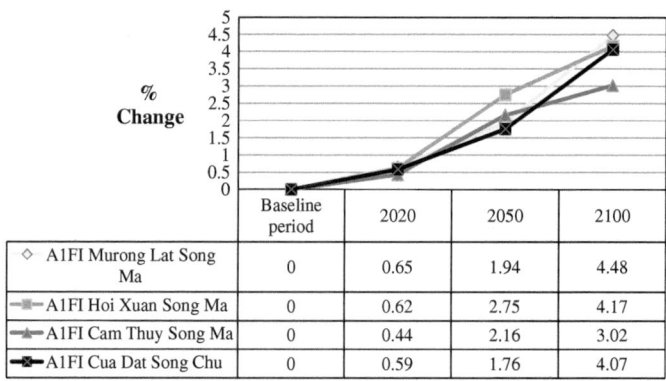

	Baseline period	2020	2050	2100
◇ A1FI Murong Lat Song Ma	0	0.65	1.94	4.48
━■━ A1FI Hoi Xuan Song Ma	0	0.62	2.75	4.17
━▲━ A1FI Cam Thuy Song Ma	0	0.44	2.16	3.02
━✖━ A1FI Cua Dat Song Chu	0	0.59	1.76	4.07

Fig. 3.11 Change in average flow in rainy season (%) at monitoring stations of river Ma and river Chu under the scenario A1FI from 2020 to 2100 compared to the baseline period (1980–2009)

The highest average monthly flows at gauging stations on the main rivers also tend to increase. Specifically, the highest average monthly flow (September) at the Muong Lat station increase by over 1.8 % by 2020 in both scenarios B2 and A1FI, further increase by 4 % by 2050 under both scenarios, and rise over 7 % under the scenario B2 and 11 % under the scenario A1FI by 2100.

Average flows in dry seasons
Dry seasons in the Thanh Hoa province occur from December to May. Different from rainy season, average flows in dry season reduce gradually in the 21st century, with a slight decrease of 0.5 % by 2020 (scenarios B2 and A1FI), a decrease of 0.7 to over 1.2 % by 2050 (scenarios B2 and A1FI) and further 3 % by 2100 according to the scenario A1FI (Figs. 3.12 and 3.13).

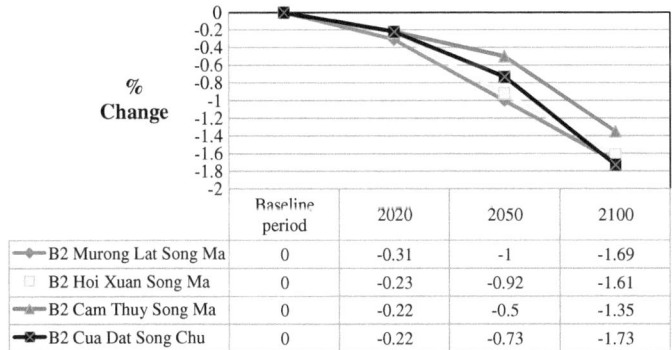

Fig. 3.12 Change in average flow in dry season (%) at monitoring stations of river Ma and River Chu under scenario B2 from 2020 to 2100 compared to the baseline period (1980–2009)

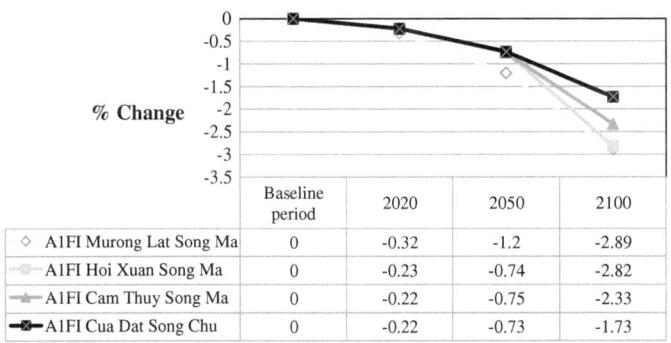

Fig. 3.13 Change in average flow in dry season (%) at monitoring stations of river Ma and river Chu under scenario A1FI from 2020 to 2100 compared to the baseline period (1980–2009)

The three driest months on the basin of the Thanh Hoa province's rivers are from February to April, while March is the driest month of the year. The flows of the three driest months tend to reduce significantly in the upper Ma river region, by nearly 4 % by 2100 at the Muong Lat station under the medium emission scenario B2 and by 6 % under the high emission scenario A1FI.

Climate change has had obvious impacts on the flows of Ma and Chu rivers. The water flow in dry season has increasingly decreased, while it has increased in the rainy seasons. Especially in the Thanh Hoa province, the flood hazard during rainy seasons has increased in the past decades.

Impacts of storm surges

According to statistical data of the past 52 years, the Thanh Hoa province has suffered direct impacts of over 100 tropical storms and tropical depressions (Source: Thanh Hoa provincial committee for flood and storm control *Ủy ban phòng chống thiên tai Thanh Hóa*). In average annually 2.4 storms make landfall or otherwise impact the Thanh Hoa province with wind speeds between the levels 8 and 12 of the Beaufort scale.

In the history of the Thanh Hoa province, there have been over 45 incidents of dyke breaches on major rivers and 13 levee breaches on smaller rivers. Strong storms have occurred in 1927, 1944, 1962, 1973, 1980, 1996, 2005 and 2007. In 2005, the Thanh Hoa province suffered from four storms, five river floods and a sweeping flood in Quan Hoa district, all of which caused serious damages to the province's regions. The two storms number 6 and 7 (in Vietnam tropical storms are numbered) arrived in Thanh Hoa between 19/9 and 27/9/2005, with wind gusts up to level 11, accompanied by extremely heavy rain. In 2007, the storm number 5 lead to heavy rain showers and storm surges, causing extremely huge floods in all rivers. The floods lead to overflows and breaches of many river dykes and resulting in serious damage.

In September 2013, the storm number 10 affected the Thanh Hoa province significantly, and resulting floods broke three river dams in the Tinh Gia district. In addition, in the Truc Lam commune there was a heavy landslide. The southern part of the Cau Tay levee of the Truc Lam commune was breached 20 m in length. Thousands of households were flooded between 1 and 1.5 m. There were landslides and deep floods along many transport roads. Also the national highway 1A from Xuan Lam to Truong Lam was flooded up to 1 m in depth, and landslides interrupted the traffic flow. Many remote hamlets were flooded and cut-off.

Storms and floods have caused severe damages in the Thanh Hoa province. In addition, they have caused many difficulties in terms of living conditions and environmental sanitation, posing danger of epidemics for the residents. Salt water from storm surges has penetrated thousands of agricultural fields, leading to soil salinisation.

Impacts of climate change on water use demand
Water demand of the studied regions is calculated taking into account water use demand for agriculture (irrigation, animal husbandry, fishery), industry and water storage maintenance.

Based on features of natural conditions, topography, geology, hydrology, climate, river systems and administrative borders, the Ma river basin is divided into 10 sub-basins:

- Sub-basin I: Upstream area of the Mã river basin
- Sub-basin II: Mộc Châu–Mường Lát districts
- Sub-basin III: Quan Hoá, Quan Sơn and Mai Châu districts.
- Sub-basin IV: Bưởi river basin
- Sub-basin V: North of Mã river basin
- Sub-basin VI: parts of the Ngọc Lạc and Thọ Xuân districts and Thiệu Hoá, Yên Định districts.
- Sub-basin VII: Bá Thước and Cẩm Thủy districts.
- Sub-basin VIII: Am river basin 11
- Sub-basin IX: Chu river basin
- Sub-basin X: South of Chu river, North of Tĩnh Gia

Water balance calculation results show that if the percentage ratio of water demand and availability is smaller than 100 % then water in that region is redundant, if the ratio becomes closer to 100 %, the region can be put in the status of water shortage alerts, and if the ratio is higher than 100 %, the region seriously lacks water (see Tables 3.1 and 3.2). Three sub-basins are considered to face water shortage already today and in the near future. The total area of the Sub-basin IX is about 200,000 ha from which about 6,000 ha are cultivated. The water shortage is highest in this sub-basin. The Sub-basin VI experiences shortage of water in some places. In the Sub-basin X, the water resources are currently sufficient to cover water demand but by 2020 there are signs of water shortage. Water recharge is higher than water demand in the sub-basins located in the upper stream of the Ma river (I–V).

Estimated water demand is based on socio-economic development master plans for 2020. There are uncertainties for the figures presented for 2050 (Table 3.3) and for 2100 (Table 3.4). According to the water balance estimations for 2050 and 2100, only the sub-basin IX will suffer from water shortage by 2100.

Table 3.1 Current status of water balance in the sub-basins of the Ma river

Sub-basin	Water demand (10^6 m^3)	Water recharge (10^6 m^3)	Ratio of demand and availability (%)
I	308.33	7,341.10	4.20
II	36.10	445.69	8.10
III	5.52	7.54	73.30
IV	1.97	6.08	32.40
V	11.20	319.94	3.50
VI	28.15	38.57	73.00
VII	145.91	550.62	26.50
VIII	57.51	1,223.64	4.70
IX	91.21	50.14	181.90
X	695.46	694.08	100.20

Table 3.2 Estimated water balance in the sub-basins of the Ma river by 2020 based on scenario A1FI

Sub-basin	Water demand (10^6 m^3)	Water recharge (10^6 m^3)	Ratio of demand and availability (%)
I	355.793	5,647.508	6.30
II	38.885	277.750	14.00
III	0.055	0.059	93.20
IV	0.009	0.020	45.00
V	0.113	1.784	6.30
VI	0.272	0.256	106.70
VII	150.916	246.193	61.30
VIII	54.113	751.575	7.20
IX	93.135	25.943	359.00
X	727.019	411.676	176.60

Table 3.3 Estimated water balance in the sub-basins of the Ma River by 2050 based on scenario A1FI

Sub-basin	Water demand (10^6 m^3)	Water recharge (10^6 m^3)	Ratio of demand and availability (%)
I	355.793	11,577.391	3.1
II	38.885	569.387	6.8
III	0.055	0.094	59.0
IV	0.009	0.031	28.5
V	0.113	2.819	4.0
VI	0.272	0.496	55.0
VII	150.916	477.613	31.6
VIII	54.113	1,458.056	3.7
IX	93.135	44.622	208.7
X	727.019	708.082	102.7

Table 3.4 Estimated water balance in the sub-basins of the Ma River by 2100 based on scenario A1FI

Sub-basin	Water demand (10^6 m^3)	Water recharge (10^6 m^3)	Ratio of demand and availability (%)
I	355.793	19,540.377	1.82
II	38.885	961.015	4.05
III	0.055	0.185	29.97
IV	0.009	0.061	14.47
V	0.113	5.549	2.03
VI	0.272	0.606	45.02
VII	150.916	583.476	25.86
VIII	54.113	1,781.233	3.04
IX	93.135	69.527	133.96
X	727.019	1,103.290	65.90

Impacts of climate change on saline intrusion

Calculations of sea level rise induced saline intrusion in the Thanh Hoa province's river system show that the salt levels vary between the estuaries of Lach Trao, Lach Sung and Lach Truong as tidal movements play an important role.

At the moment, water salinity of 4 ‰ has penetrated about 26.3 km upstream into Ma River, about 22.8 km into Len river, and into the whole length of Lach river. Under sea level rise scenario A1FI for 2020, salt water penetrates about 28 km into the Ma River and 24 km for Len River. Based on sea level rise scenario A1FI for 2050, salt water penetrates about 31.5 km into the Ma River and 27.5 km into the Len River. In 2100 (A1FI), salt water penetrates about 34.5 km into Ma river and almost completely into the Len river.

Climate change and sea level rise obviously have strong impacts on saline intrusion into river basins of the Thanh Hoa province, and thus substantially affect the water use in all sectors of the provinces (Table 3.5).

Table 3.5 Penetration length of saline intrusion for 2020, 2050 and 2100 in rivers Ma, Len and Lach Truong based on sea level rise scenarios B2 and A1FI

Scenario/ year	Ma river			Len river			Lach Truong river	
	Saline intrusion (km)	Salinity 4 ‰		Saline intrusion (km)	Salinity 4 ‰		Saline intrusion (km)	Salinity 28 ‰
		Peak flow (km)	Low flow (km)		Peak flow (km)	Low flow (km)		
Current status	26.3	17.5	14	22.8	17.8	15.3	River wide	6
B2/2020	28	19.5	15.5	24	17.4	16	River wide	7.0
A1FI/2020	28	19.8	16	24	18	16	River wide	7.5
B2/2050	31	21.8	20	27	18.5	18	River wide	7.8
A1FI/2050	31.5	22.5	21.5	27.5	18.8	18.5	River wide	7.8
B2/2100	34	25	23	River wide	20.3	19.5	River wide	7.8
A1FI/2100	34.5	25.5	24	River wide	20.5	19.7	River wide	8

3.6 Impacts of Socio-economic Development and Climate Change on Groundwater

Socio-economic activities can have substantial impacts on groundwater quality and quantity. These include urbanization processes, industrialization, farming activities, services and tourism.

Urbanization process
Urbanization leads to increase in population and population density, raising the demand for use of water resources in general, and groundwater in particular. Urban development processes also lead to increase in solid waste and domestic wastewater.

Industrialization process
Industrial development is accompanied by the increase in water consumption. Insufficient environmental standards and low restrictions on the use of natural resources quickly lead to unsustainable exploitation of resources, as well as contamination of the environment (especially soil and water resources).

Agriculture and rural area development
The lack of control of the use of pesticides, manures and fertilizers leads to the contamination and pollution of surface water and groundwater resources, causing danger to species and human beings directly through water sources and indirectly through food chain.

Sewage and wastes from animal husbandry, as well as domestic waste and wastewater in rural areas are usually not collected and treated to ensure environmental sanitation, which poses substantial danger to surface water and groundwater.

Services and tourism development

Wastes from tourism activities which are not controlled and treated and poses a potential risk to lead to environmental pollution. The development of tourism is currently rather spontaneous and often incompliant with the master plans. This uncontrolled development leads to unsustainable exploitation and contamination of natural resources and ultimately the destruction of the ecological environment surrounding tourist destinations.

Sanitation facilities and wastes

Sanitation facilities close to water supply facilities (over 50 % of water supply facilities are less than 10 m away from sanitation facilities) pose highly potential risk for groundwater pollution.

The amount of domestic wastes and wastewater are increasing, while sewerage and waste treatment infrastructure in rural areas are rare. Both dispersed wastes and wastewater possess substantial threats to the environment.

Domestic and industrial wastes are being dumped daily at the river banks and lakes, on the back of the villages, on mountains, along the coast, in road junctions or at the edge of the fields. By this dispersion, there is a danger that individual houses are clean while common places are dirty, thus directly affecting groundwater and surface water resources. Untreated wastes from industrial production and processing activities are further sources of water resources pollution.

Exploitation of groundwater

Results from studies show that domestic water sources commonly used by peoples living along the coasta of the Thanh Hoa province are private dug wells or drilled wells.

Two centralised water supply facilities are in operation in the province, in But Son town and Tĩnh Gia town, supplying water to about 300 households. Those households without water supply facilities collect rain water, fetch water from rivers and/or lakes, or transport water from long distance in dry seasons.

Currently, there are 563,950 facilities which exploit groundwater in the whole province, of which 218,774 are drilled wells (mostly drilled wells of UNICEF type) and 345,476 are dug wells with a total exploitation capacity of up to 282,519 m^3/day (Trung 2005a, b).

Impacts of socio-economic activities on groundwater resources

The impacts of socio-economic activities on groundwater resources can be severe and affect both groundwater quality and quantity.

The uncontrolled seepage from landfills or wastewater infiltration may worsen the quality of groundwater. In addition, salt water may infiltrate the aquifer through direct penetration, or through drilled wells and dug wells. The latter one may also lead to further contamination by NH_4, salts and microorganisms.

Accurate assessments of groundwater use are difficult because several wells are drilled without centralized registration. An additional factor is the failure of complying with regulations for the protection of aquifers.

Development and operation of landfills that do not meet the existing standards can unintentionally damage the natural protection layer, which increases the exposure of groundwater to contamination.

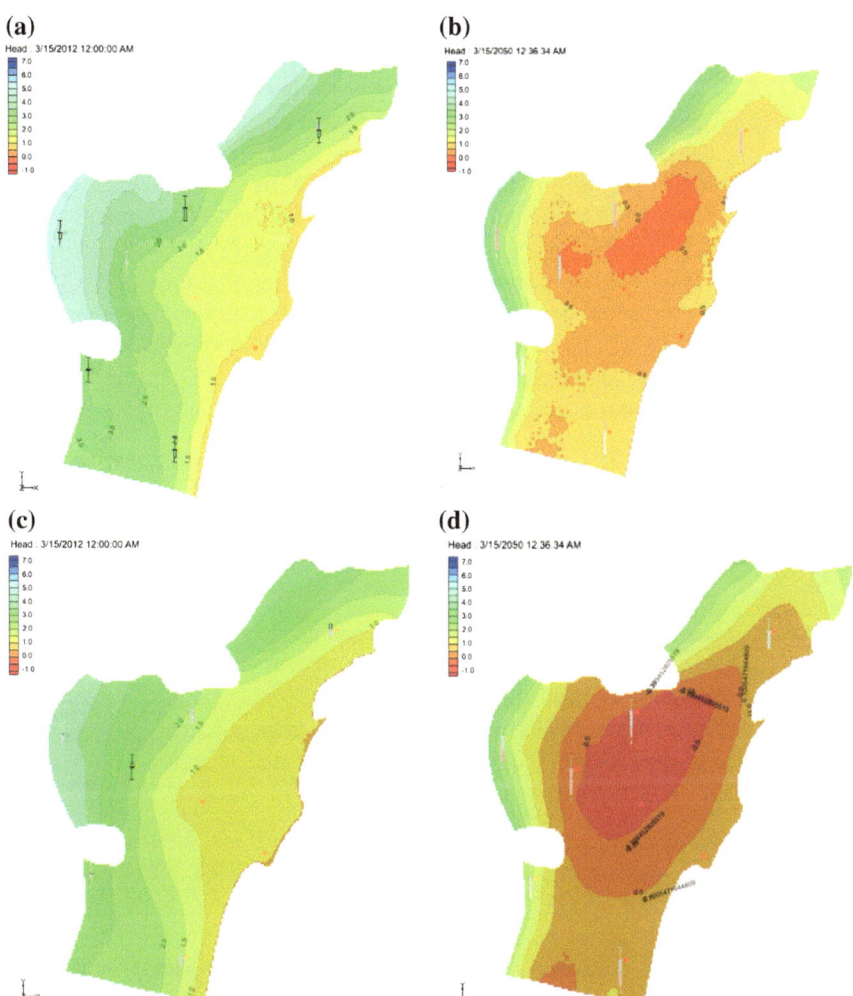

Fig. 3.14 Recession of groundwater levels during dry seasons due to increased water exploitation in 2012 and 2050 in two aquifers of the Thanh Hoa province, Holocene (qh) and Pleistocene (qp). The projected changes are based on the climate change scenario B2 and estimated population increase. **a** Water level of the qh aquifer, dry season in 2012. **b** Water level of the qh aquifer, dry season in 2050. **c** Water level of the qp aquifer, dry season in 2012. **d** Water level of the qp aquifer, dry season in 2050

Excessive exploitation and lowering of water levels may cause hazardous substances to be released into the groundwater. Substantial groundwater (drinking water) contamination has been reported to the project, caused by domestic local waste, untreated sewage and graveyards, etc.

The projections of water levels in coastal regions under both climate change and socio-economic development scenarios show that with the current rate of exploitation (about 120,000 m³/day), the deepest level of the Holocene (qh) aquifer is 1.9 m (Quảng Chính–Quảng Xương) and that of the Pleistocene (qp) aquifer 1.4 m (Nga Hưng—Nga Sơn). The increase in population leads to a rise in water consumption. The population by 2050 will reach about 2,938,349 persons, and the demand for water use will be about 295,835 m³/person (100 L/person/day). Exploiting 85,000 m³/day from which 65 % from groundwater sources may lower the deepest ground water level of the qh aquifer to 0.56 m and the qp aquifer to −0.54 m below mean sea level. Respectively, by 2100, the estimated population will be 3,981,000 persons. Water demand is approximated to be 100 L/person/day, and total water demand 399,000 m³/day. It is expected that about 60 % of the water demand will be on groundwater sources (241,960 m³/day). Based on these scenarios, it is assumed that the deepest groundwater level of qh aquifer will be 0.03 m above mean sea level and qp aquifer −1.15 m below mean sea level (see Fig. 3.14).

3.7 Scenarios on Changes to Groundwater Resources

During the period 1990–2011, groundwater recharge decreased by 15 % in dry seasons and in rainy seasons by 13 %. This is due to rising temperatures and increasing evaporation, as well as groundwater extraction. In order to estimate the possible scarcity of future water resources, climate change scenarios were combined with the groundwater exploiting scenarios in the coastal regions.

Scenario 1
Scenario 1 is based on the high (A1FI) emission scenarios for 2050 and 2100 and water withdrawal increase due to a rise in water demand which is based on prognosis of the population growth.

By 2050, the population of the Thanh Hoa coastal area is expected to reach 2,938,000. The water demand is approximated to be 100 L/person/day, leading the total water usage demand to be 293,835 m³/day. Approximately 50 % of the water demand will be met by groundwater resources (146,917 m³/day), and the remaining demand will be met by surface water.

By 2100, the aquifer will be diminished by 65 km² due to sea level rise of 86 cm according to the highest emission scenario (A1FI). Thanh Hoa's coastal population is calculated to be of 3,981,000 in 2100. The water reaches up to 100 L/person/day, leading the total water demand to be 399,000 m³/day. About 60 % of the water demand will be met by groundwater resources (241,960 m³/day), and the remaining demand will be met by surface water (see Fig. 3.15).

Scenario 2
Scenario 2 uses the medium emission scenario B2 for 2,050 m and both unchanged and increased water exploitation which is based on population growth prognosis. Reductions of dynamic groundwater reserves are compared to the current status (Table 3.6).

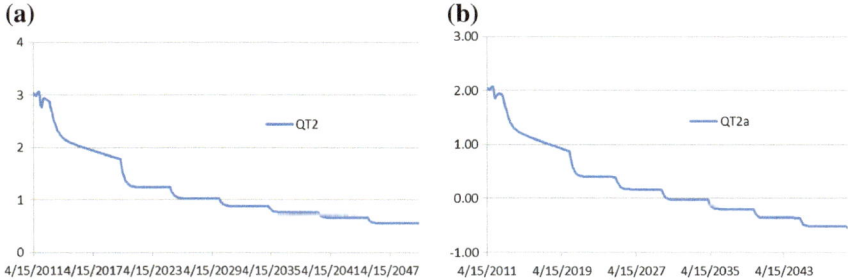

Fig. 3.15 Possible recession of groundwater levels in the Thanh Hoa coastal area due to increasing exploitation by 2050 in two aquifers, Holocene (qh) and Pleistocene (qp). **a** Water level (m asl) of the Holocene (qh) aquifer in 2011–2050. **b** Water level (m asl) of the Pleistocene (qp) aquifer in 2011–2050

Table 3.6 Results of projections run for 2050 for the Holocene (qh) aquifer in the Thanh Hoa coastal area

Time	Recharge volume (m³/day)	Exploitation volume (m³/day)	Dynamic reserve of qh aquifer (m³/day)
March (dry season), 2012	478,157	52,240	140,835
Dry season 2050, climate change (B2) and increased exploitation	439,167	80,186	138,242
Dry season 2050, climate change (B2) and unchanged exploitation	439,167	52,240	135,451

Scenario 3

Sea level rises by 86 cm according to the high emission scenario A1FI and the volume of exploitation increases due to increased water demand, which is based on prognosis of the population growth.

By 2100, Thanh Hoa's coastal population might reach 3,981,000. Water demand is approximated to be 100 L/person/day, leading the total water demand to be 399,000 m³/day. Sixty percent of the water demand will be met by groundwater resources (241,960 m³/day), and the remaining demand will be met from surface water. Areas of Holocene and Pleistocene aquifers are diminished by 65 km² owing to the sea level rise.

The results show that water level will be lowered in both aquifers, and stronger in the Pleistocene (qp) aquifer, with the deepest level of −1.4 m bsl (Hoang Trung, Hoang Hoa district) (Tables 3.7 and 3.8).

It is difficult to separate climate change and socio-economic impacts on groundwater. Studies on the change in water quality with TDS values at medium emission scenario B2 and high emission scenario A1FI for sea water level rise for 2050 and 2100 have been carried out.

Table 3.7 Results of projections run by 2100 for Holocene (qh) aquifer in the Thanh Hoa coastal area

Time	Exploitation volume (m³/pers.)	Water level (m asl)				
		QT2	QT6	QT11	QT12	QT13
March (dry season), 2012	52,240	2.83	4.83	3.04	1.90	2.13
Dry season, 2050	80,186	0.59	2.30	1.46	0.83	0.71
Dry season, 2100	105,010	0.24	1.14	0.47	0.83	0.62

Table 3.8 Results of projections run by 210, for Pleistocene (qp) aquifer in the Thanh Hoa coastal area

Time	Exploitation volume (m³/pers.)	Water level (m asl)					
		QT2a	QT6a	QT11a	QT12a	QT13a	QT7a
March (dry season), 2012	66,035	2.53	4.27	2.84	2.65	2.72	2.90
Dry season, 2050	98,901	−0.48	1.79	1.30	0.49	0.44	1.30
Dry season, 2100	128,577	−1.04	0.31	0.07	0.40	0.25	−1.06

The results show that when the water level drops, the change in quality will occur. These changes are demonstrated with three scenarios.

Scenario 4

Scenario 4 uses the medium emission scenario for 2050. For the Holocene (qh) aquifer, in the areas with TDS < 1.5 g/l, i.e. namely fresh or brackish water, changes are quite obvious. Since it is the topmost layer, the groundwater could easily be affected by saline intrusion due to sea water rise during flood events.

For the Pleistocene (qp) aquifer, the area of fresh water regions is diminished, however, the change is not considerable (see Fig. 3.15, item c).

Scenario 5

Scenario 5 is calculates with a sea level rise for 2100 derived from the medium emission scenario B2 for 2100.

For the Holocene (qh) aquifer, in the area with TDS < 1.5 g/l, namely fresh and brackish water, changes are quite obvious. Since it is the topmost layer, the groundwater could easily be affected by sea water, while saline intrusion due to storm surges events will increase as compared to the period of 2050 (see Fig. 3.16). In some places adjacent to the sea, the TDS content will have high increase (see Figs. 3.17 and 3.18).

For the Pleistocene (qp) aquifer, the area of fresh water will gradually be diminished, however, not as much as today. In the places adjacent to the sea the content of TDS gradually increases. In the places further away from the sea shore, TDS values gradually decrease (see Fig. 3.19).

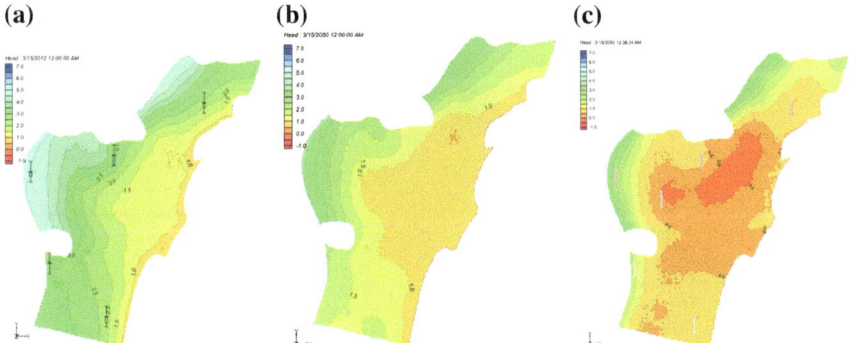

Fig. 3.16 Impacts of climate change (B2 scenario) on groundwater level of the Holocene (qh) aquifer in the Thanh Hoa coastal area. **a** Dry season by 2012, water demand 52,240 m³/pers, dynamic water reserve 140,835 m³/day. **b** Dry season by 2050, water demand 52,240 m³/pers, dynamic water reserve 135,451 m³/day. **c** Dry season by 2050, water demand 80,186 m³/pers, dynamic water reserve 138,242 m³/day

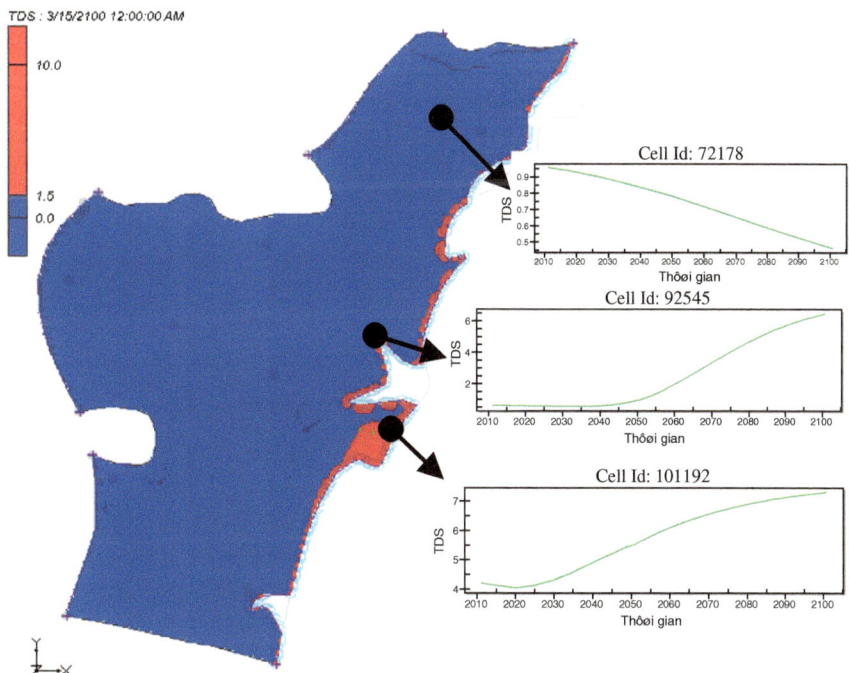

Fig. 3.17 TDS content of the Holocene (qh) aquifer in Thanh Hoa coastal area from 2010 to 2100. The map indicates the modeled TDS content in 2100

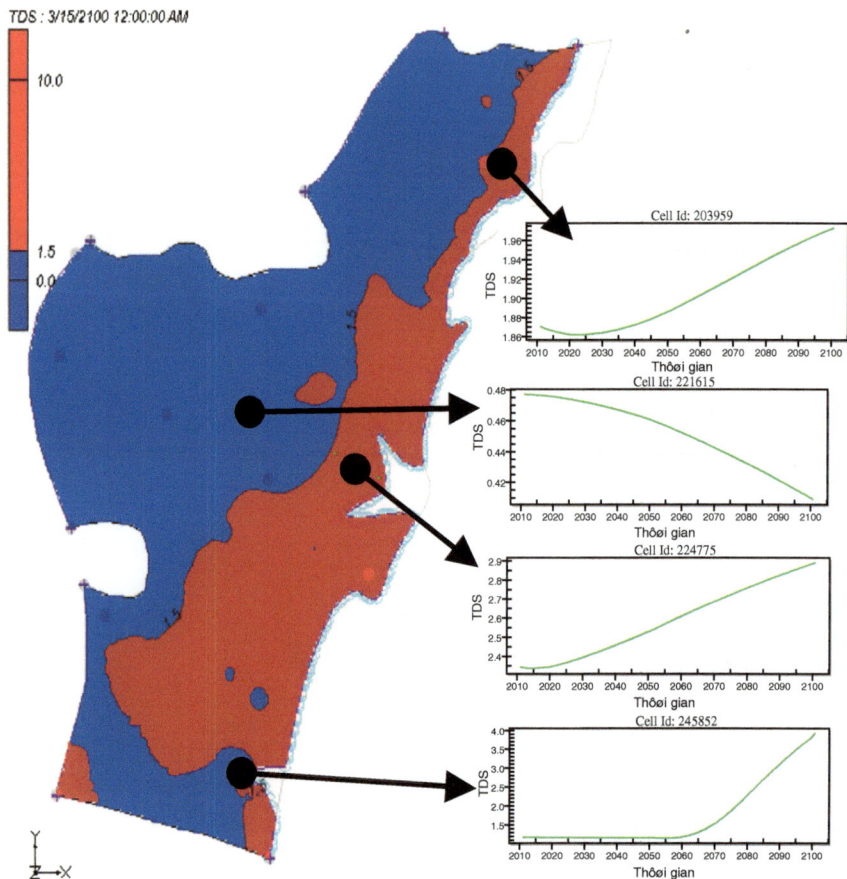

Fig. 3.18 TDS content of the Pleistocene (qp) aquifer in Thanh Hoa coastal area from 2010 to 2100. The map indicates the modeled TDS content in 2100

Scenario 6

Scenario 6 is based on the medium emission scenario B2 for 2100. By then, Thanh Hoa's coastal population reaches up to 3.981.000. The water demand is 100 L/person/day, leading to a total of 399.000 m^3/day. About 16 % of water demand will be met by groundwater resources (44,870 m^3/day), and the remaining water demand will be met from surface water sources. Area of the Quaternary aquifers is diminished by 65 km^2 due to sea level rise.

The results show that under these assumptions the groundwater water level is lowered to a lesser degree. Because the groundwater flows mainly runs from north-west to south-east, the Holocene (qh) aquifer will be recharged naturally—if exploitation decreases. The scenario 6 was the scenario chosen for further analysis with the stakeholders.

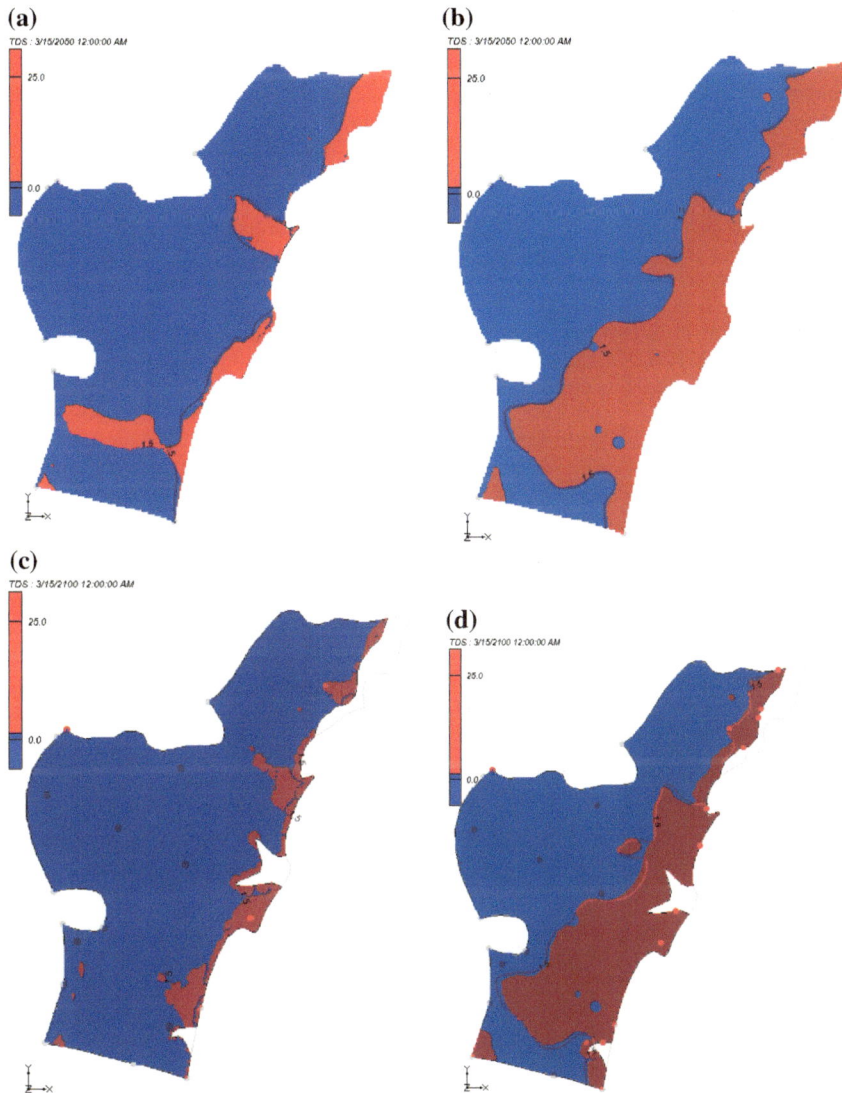

Fig. 3.19 TDS content in the Holocene (qh) and Pleistocene (qp) aquifers of the Thanh Hoa coastal area in 2050 and 2100 based on sea level rise scenarios. Salt water intrusion marked in *red*. **a** TDS qh aquifer 3/2050. **b** TDS qp aquifer 3/2050. **c** TDS qh aquifer 3/2100. **d** TDS qp aquifer 3/2100

References

Cục thống kê tỉnh Thanh Hóa (2010) Niên giám thống kê tỉnh, Thanh Hóa (Thanh Hoa Statistics Office (2010) Thanh Hoa statistical yearbook, Thanh Hoa)

MONRE (Bộ Tài Nguyên Môi Trường) (2012) Kịch bản Biến đổi khí hậu, nước biển dâng cho Việt Nam. Hà Nội (Ministry of Natural Resources and Environment (2012) Climate change scenarios and sea level rise for Vietnam. Hanoi)

Nguyễn VT và nhóm nghiên cứu (2010) Biến đổi khí hậu và tác động ở Việt Nam. Viện khoa học Khí tượng Thủy văn và Môi trường, Hà Nội (Nguyen VT et al (2010) Climate change and impacts in Vietnam. Scientific Institute of Meteorology, Hydrology and Environment, Hanoi)

Trung tâm nước sinh hoạt và Vệ sinh môi trường nông thôn (2005a) Dự án điều tra quy hoạch khai thác nguồn nước phục vụ yêu cầu cấp nước sinh hoạt và phát triển kinh tế xã hội vùng ven biển Thanh Hóa đến năm 2010, định hướng đến năm 2020, Thanh Hóa (Center for Potable Water and Rural Sanitation (2005) Investigation of groundwater extraction to meet the requirements of domestic water supply and socio-economic development of the coastal areas of Thanh Hoa to 2010 and orientations towards 2020, Than Hoa)

Trung tâm nước sinh hoạt và Vệ sinh môi trường nông thôn (2005b) Dự án điều tra quy hoạch nguồn nước phục vụ cấp nước sinh hoạt và phát triển KT-XH vùng ven biển Thanh Hóa đến năm 2015, Thanh Hóa (Center for Potable Water and Rural Sanitation (2005) Investigation on water planning water supply serving and socio-economic development of the coastal areas of Thanh Hoa in 2015, Thanh Hoa)

Trung Tâm Quan trắc và Dự báo Tài nguyên nước (2010) Tài liệu quan trắc tài nguyên NDĐ mạng quan trắc Quốc gia năm 2011–2012, Hà Nội (Center for Monitoring and Forecast of Water Resources (2010) Resources and monitoring network for national groundwater monitoring 2011–2012, Hanoi)

Chapter 4
Impacts of Climate Change on the Ba Ria–Vung Tau Province

Abstract The Ba Ria–Vung Tau province is both an important industrial area and tourism hot spot. The development of these two sectors has led to substantial stresses to the environment, especially water quality and quantity. The socio-economic development remains at a high pace. The natural salinization of aquifers might become more important due to future increase of groundwater demand. Detailed water resource and balances were started, also taking climate change impacts and socio-development impacts into account. These form the basis to build future development and impact scenarios to develop and implement climate change adaptation measures in close cooperation with local stakeholders.

The Ba Ria–Vung Tau province is located in the South East Region, belonging to the southern economic region. The geographical coordinates are: 107°00′01″–107°34′18″ eastern longitude and 10°19′08″–10°48′39″ northern latitude. The province has a total area of 198.95 ha, accounting for 0.6 % of Vietnam and 8.3 % of the South East Region area. With a population of 996.879 it accounts for 0.95 % of the national population, and the population density of 509 persons/km^2 equals 1.95 times the national population density.

With regards to administration, the Ba Ria–Vung Tau province has eight administrative units, including: Vung Tau city, Ba Ria city and districts of Chau Duc, Xuyen Moc, Tan Thanh, Long Dien, Dat Do and Con Dao islands, with total 82 communes, villages and towns (Fig. 4.1).

4.1 State of the Environment in the Ba Ria–Vung Tau Province

The ambient air quality in areas such as Tan Lam commune (Xuyen Moc), Suoi Rao commune (Chau Duc) and Bai Truoc and Bai Sau tourism areas is fairly good. All monitored air quality indicators such as suspended dust, noise, SO_2, NO_2 and CO are

© The Author(s) 2015
P. Schmidt-Thomé et al., *Climate Change Adaptation Measures in Vietnam*, SpringerBriefs in Earth Sciences, DOI 10.1007/978-3-319-12346-2_4

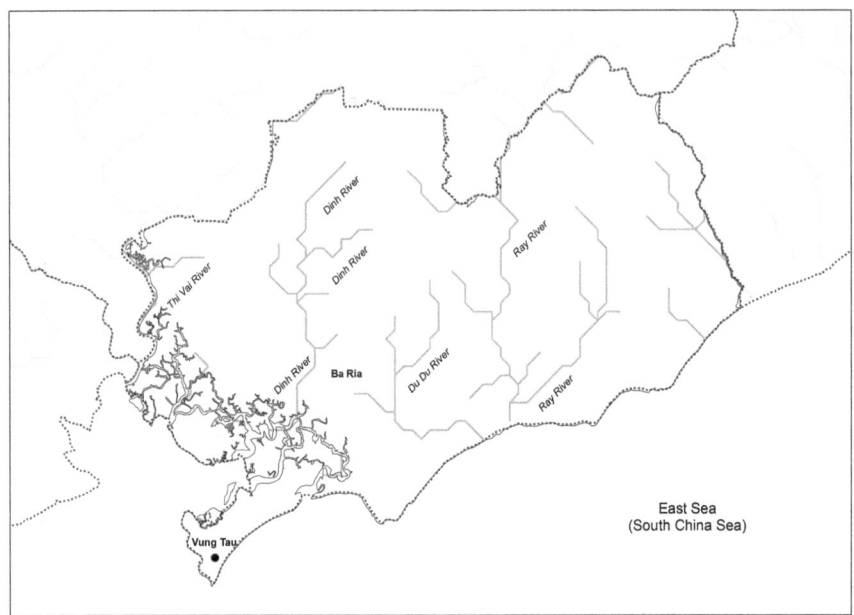

Fig. 4.1 Map of Ba Ria–Vung Tau

within the accepted air quality limits specified in the Vietnam's National Technical Regulations (QCVN 05: 2013/BTNMT). However, air quality in the fishing villages are generally polluted with SO_2 (fishing villages of Loc An and Ben Dinh), NH_3 (Loc An fishing village), and H_2S (Hoi Bai fishing village). The air quality in the Toc Tien dumping area (Tan Thanh) is frequently highly polluted by H_2S.

The air quality in some locations in urban areas experience noise and SO_2 pollution, but other monitoring indicators are all within accepted limit values pursuant to the Vietnam standards on environment (QCVN 05: 2013/BTNMT).

Water quality of lakes mostly meets the national technical regulations on surface water quality (QCVN 08: 2008/BTNMT). Only some lakes exceeded permitted limit values of certain monitored elements, such as total suspended solids (TSS), organic matter (DO, COD, BOD_5), nutrients (NH_4^+, NO_2^-) and microorganisms.

The water of rivers at monitoring stations is often polluted. The types and levels of pollution depend on locations of monitoring points and have seasonal variation. The exceeded limit values mainly consist of total suspended solids (TSS), organic matter (COD, BOD_5), nutrients (NH_4^+, NO_2^-, PO_4^{3-}), iron and microorganism. Other monitored indicators all meet permitted limit values indicated in the Vietnam's national technical regulations (QCVN 08: 2008/BTNMT).

Coastal seawater is polluted by oil products at some monitoring stations. The content of micro organism in seawater at some monitoring points is also high (Nguyễn et al. 2010).

4.2 Hydrogeology

Holocene porous aquifer (qh)

The Holocene porous aquifer (hereinafter referred to as qh aquifer) is formed by coarse-grained soil and rocks, and Holocene sediments of which most notable are eolic sediments that formed long sand dunes along the beaches. The lithologic composition comprises mainly loose fine-grained sand, sand mixed with silt and white or light grey silty sand. Thickness varies from several meters (divergent boundary) up to 22 m. The aquifer is interrupted by bedrock formations and older water bearing formations. The aquifer is covered by the poor water bearing Holocene formation Q_2 and lies on top of the very poor water bearing upper Pleistocene formation Q_1^{2-3}. The total area of the qh aquifer is about 700 km^2.

Results of pumping tests at several monitoring wells have shown that the qh aquifer has a poor to medium productivity. With the pumping rate (Q) 0.1–1.1 L/s (average 0.6 L/s) the drawdown (S) is of 0.30–5.1 m (average 2.35 m) and the specific discharge (q) is 0.1–0.6 (L/s)/m (average 0.2 (L/s)/m). The water level depending on the geological conditions of the aquifer varies in the range of 0.5–8.9 m with the average thickness of 2.66 m.

Fresh water is found on high terrains or the centres of sand dunes. Analysis of chemical composition of water have brought the following results: total dissolved solids (TDS) 0.04–0.1 g/l (average 0.4 g/l), low pH of weak acid to neutral ranging within 3.8–8.5 (average 7.1), water hardness varying from very soft to hard water within the range of 0.2–7.1 meq/l (average 2.3 meq/l).

Nitrogen compounds are commonly found with high levels. Particularly the frequent presence of NH_4^+ indicates that water is polluted from surface sources (fertilizers, domestic sewage and industrial wastewater, etc.). The contents in many places exceed the groundwater quality limit values (QCVN 09: 2008/BTNMT). Saline water is often found in low terrain or on the fringe of the sand dunes. The total area of saline water is about 30 km^2. Analysis of chemical composition of water brought the following results: total dissolved solids (TDS) 1.1–4.7 g/l (average 2.0 g/l), slightly neutral water with pH varying within the range of 7.0–8.0 (average 7.6), water hardness varying from very soft to hard water within the range of 0.1–0.4 meq/l (average 0.2 meq/l).

Pleistocene porous aquifer (qp)

The Pleistocene porous aquifer (hereinafter referred to as qp aquifer) is formed of coarse-grained soil and rocks of the Cu Chi (Q_1^3cc) and the Thu Duc ($Q_1^{2-3}t$đ) formations. The lithologic composition comprises mainly of weakly consolidated or loose fine-grained sand and light grey silty sand. The aquifer is not continuous due to bedrock intercalations and older exposed water bearing formations. The total area is approximately 233.5 km^2. The qp aquifer is normally covered by very poor water bearing formation $Q_1^{2-3}t$đ and lies on top of very poor water bearing formation Q_1^1đc. Thickness found at the boreholes varies between 1.6 and 20.0 m with the average thickness of 13.2 m.

Results of pumping tests at several monitoring wells show that this is a poor to medium water bearing aquifer. With the pumping rate (Q) 0.2–6.3 L/s (average 2.09 L/s), drawdown (S) is 1.5–19.3 m (average 8.2 m) and specific discharge (q) is 0.04–1.4 (L/s)/m [average 0.34 (L/s)/m]. The water level, depending on the geological conditions of the aquifer, varies in the range of 0.01–13.00 m with the average water level of 4.0 m.

Ultra fresh water is generally found in rising terrains or mountain sides with rainwater as the main recharge source. The chemical composition of the fresh water area is following: total dissolved solids (TDS) 0.03–0.1 g/l (average 0.1 g/l), pH values range between 4.8 and 6.7 (average 5.7), very soft water with water hardness of 0.1–0.3 meq/l (average 0.1 meq/l).

Fresh water is generally found in low terrains or on the fringe of river valleys of, streams, lakes and exchanging water with surface water systems. Analysis of chemical composition of water brought the following results: total dissolved solids (TDS) 0.1–0.8 g/l (average 0.3 g/l), pH values between 3.4 and 8.1 (average 7.0), water hardness varying from very soft to slightly hard water within the range of 0.4–3.3 meq/l (averaging: 1.3 meq/l).

Nitrogen compounds are commonly found in high levels, in particular with a frequent presence of NH_4^+, suggesting that water is polluted from surface sources (fertilizers, domestic sewage and industrial wastewater, etc.). The content of NH_4^+ in many places slightly exceeds the limit values. However, the contents of nitrogen compounds found in bore hole KH8 (Long Dien, Long Dat district) were rather high, NH_4^+ 0.20 mg/l, NO_3^- 25.3 mg/l and NO_2^- 2.9 mg/l while compared to corresponding limit values, 0.1, 15 and 1.0 mg/l.

Saline water is commonly distributed in low terrain or on the fringe of valleys of rivers, streams, lakes in southern coastal plains with an area of about 23.5 km^2. Analysis of chemical composition of water brought the following results: total dissolved solids (TDS) 1.4–35.5 g/l (average 19.3 g/l), pH varying in the range of 3.2–8.6 (average 7.0), very hard water with hardness varying in the range of 6.9–125.0 meq/l (average 71.2 meq/l).

Mid Pliocene porous aquifer (n_2^2)

Mid Pliocene porous aquifer (hereinafter referred to as n_2^2 aquifer) is formed of coarse-grained soil and rocks, located below the Ba Mieu (N_2^2bm) formation and the Suoi Tam Bo (N_2^2stb) formation. The total area is 175.6 km^2. The lithologic composition comprises mainly sand, pebbles, gravels, coarse sand and silty sand. The aquifer thickness varies from 2.1 to 30 m, average thickness is 13.3 m. The aquifer is widely distributed and is interrupted by bedrock intercalations. The aquifer tends to go deeper towards the sea and its thickness tends to increase accordingly. It is normally covered with younger formations and is located on top of the Mesozoic aquifer.

Results of pumping test have shown that this is a poor to high water-bearing aquifer: Wit the pumping rate (Q) of 0.02–22.2 L/s, the drawdown (S) is 0.3–33.6 m and the specific discharge (q) is 0.001–7.1 (L/s)/m [average 1.4 (L/s)/m]. The static water level varies between 0.4 and 23.0 m (average 6.3 m).

Ultra fresh water is generally distributed in rising terrains or mountain sides with rainwater as the main water sources. Analysis of chemical composition of water brought the following results: total dissolved solids (TDS) 0.03–0.1 g/l (average 0.06 g/l), low acid to neutral with pH values in the range of: 4.6–7.3 (average 6.1), very soft water with water hardness in the range of 0.1–0.5 meq/l (average 0.3 meq/l).

Analyses of chemical composition of fresh water brought the following results: total dissolved solids (TDS) 0.1–0.7 g/l (average 0.2 g/l), pH ranging within 4.6–9.3 (average 7.2), very soft water to soft water with hardness varying within the range of 0.2–11.2 meq/l (average 1.4 meq/l). Nitrogen compounds are commonly found in high levels, particularly the frequent presence of NH_4^+ suggests that the water is polluted from surface sources (fertilizers, domestic sewage and industrial wastewater, etc.). The content of NH_4^+ in many places exceeds the limit values of QCVN 09: 2008/BTNMT.

Saline water is frequently distributed on the fringe of valleys of the Thi Vai river and southern coastal plains with an area of about 13.1 km^2. Analysis of chemical composition of water brought the following results: total dissolved solids (TDS) 1.1–6.5 g/l (average 4.4 l), pH varying within the range of 3.8–8.3 (average 5.3), water hardness varying from hard to very hard within the range of 5.6–168.0 meq/l (average 58.9 meq/l).

Fissured rocky Mesozoic aquifer (ms)

The fissured rocky aquifer of Paleozoic and Mesozoic age (hereinafter referred to as ms aquifer) was formed with the fissured zones on the upper most part of the La Nga (J_1ln) and the Long Binh (J_3-K_1lb) formations. The aquifer is widely distributed in the whole province of Ba Ria–Vung Tau and is only interrupted with non-water bearing formation of GDiK, with distribution area of about 264.2 km^2. The lithologic composition comprises claystone, sandstone, siltstone, andesite tuff, dacite tuff, gritstone tuff, black shale and siltstone tuff. Total thickness is quite high, but within the scope of the current study, the water bearing fracture zone is normally 80 m thick. The upper zone is typically completely weathered to form weak water bearing loose clastics. The ms aquifer is exposed as narrow strips around Long Hai and Xuyen Moc mountains. Most of the remaining area is covered with porous Cenozoic sediments.

Results of pumping test have shown that this is a poor to rich water bearing aquifers. With pumping rate (Q) of 0.2–6.9 L/s (average 2.6 L/s), the drawdown (S) is 1.58–31.4 m (average 11.7 m) and the specific discharge (q) is 0.003–0.8 (L/s)/m (average 0.2 (L/s)/m). Static water level varies within the range of 2.4–9.8 m (average is 3.2 m above mean sea level).

The chemical composition of water is: total dissolved solids (TDS) 0.04–0.5 g/l (average 0.3 g/l), pH varying from 6.5–8.5 (average 8), water hardness varying from very soft to slight hard, ranging within 0.6–4.4 meq/l (average 1.9 meq/l). In the ms aquifer, the contents of nitrogen compounds are not high and are much lower than the limit values in QCVN 09: 2008/BTNMT.

The unconfined aquifers have a hydraulic connection with the surface water systems and adjacent aquifers. The main water sources are rainwater and water leaking from adjacent aquifers. Water is mainly drained to the west and to the sea in the south or through the rivers and streams. Water is exploited by drilled wells, and there is also leakage to adjacent aquifers.

4.3 Socio-economy and Development Objectives

In the Ba Ria–Vung Tau province, the industrial production and construction account for 83.47 % of the total output value while service sector accounts for 10.1 %, agriculture and aquaculture account for 6.43 %.

Industrial production and construction
Industrial production output has increased 2.31 times in the past 5 years. Industrial production and construction account for 83.47 % of the total economic output value. Meanwhile industries such as mechanical, electrical, fertilizer, steel etc. still account for a large proportion of the total industrial output the last 5 years have seen the emergence of new industrial products as well as incremental increase in output of several existing products.

Tourism—Services
Over the past 5 years, the annual average growth of tourism in the Ba Ria–Vung Tau province has reached 14.87 %. So far, there are 56 hotels with 2,500 rooms in the entire province. Among which 46 hotels are ranked 4-stars and up with 2,000 rooms. There are also more than 400 guest houses and hostels with more than 2,300 rooms. Every year, almost 1.4 millions tourists visit Ba Ria–Vung Tau. Annual tourism revenue is VND 2.673 billion (0.1 million USD). 18 modern resorts were recently constructed along the Vung Tau–Long Hai–Xuyen Moc beaches to meet tourist demands.

Agriculture
Agriculture land accounts for 57 % of natural land. Rural population accounts for 55 % of total population of the entire province. It could be presumed that agriculture plays a critical role in the socio-economic development strategy of the Ba Ria–Vung Tau province as its annual average growth is 17 %. On the other hand cultivation and breeding sectors has inclined to decrease due to urbanization in the past few years. Currently, only about 50 ha of agriculture land are cultivated.

Forestry
In the past 5 years, average annual growth rate of forestry is 3.5 %. Afforestation increased by 3.9 % per year, logging and forest products by 2.6 %, industrial forestry activities by 3.1 % and forestry services by 6.5 % per year.

Aquaculture
The total surface water area for aquaculture is more than 7,723 ha of which fresh water aquaculture covers approximately 2,000 ha. Source: Cục thống kê Bà Rịa–Vũng Tàu (2010).

Socio-economic development goals by 2020

Annual GDP growth during the period 2011–2015 will reach 11.8 % (exclusive of oil and gas industry, which is 16.6 %) and for the period 2016–2020 is expected to be 11.1 % (exclusive of oil and gas industry, which is 13.4 %).

Economic structure promotes the development of service sector, especially tourism services, seaport services and trading. The aim is to develop advanced technologies with high labor productivity, to ensure high quality production, and to promote the transformation of structure in each economic sector to raise efficiency and aims towards reaching a knowledge-based economy. The economic structure by 2020 shall be: industrial production and construction accounts for 61.6 %; service accounts for 36.8 %; agriculture, forestry and fishery account for 1.7 % (if oil and gas industry are excluded, the figures for economic structure are 53.2; 44.8; 2 %). The average GDP per capita by 2020 shall reach 27.000 USD, which is 2.4 times higher as compared to 2010.

It is intended to develop infrastructure system in the rural areas. The aim is that by 2020, 100 % of rural households have electricity and clean water supply. Also the education system is under revision to meet higher standards (Bà Rịa–Vũng Tàu 2012; Nguyễn et al. 2010).

4.4   Impacts of Socio-economic Development and Climate Change on the Environment

Urbanization

Population growth has increased the amount of domestic sewage, most of which are only preliminarily treated in septic tanks. Hence, most of domestic sewage from residential areas is heavily polluted with organic matters (COD, BOD), suspended solids and Coliform bacteria. This poorly purified water is discharged into common sluices or channels, ponds or lakes.

The urbanization process increases the amount of solid wastes. Currently, the Ba Ria–Vung Tau province produces about 1,000 tons of domestic wastes per day. Solid wastes in Vung Tau city, Ba Ria town and Tan Thanh district (accounts for 70 % of solid wastes of the entire province) is collected and transported to the Toc Tien landfill, located in the central part of the province. Domestic wastes from several districts are deposited in temporary landfills, each of which has an area of 1–2 ha. These temporary landfill areas have become overloaded and may cause environmental pollution (Nguyễn et al. 2010).

Industrialization

With fast socio-economic development, especially in industrial production, the Ba Ria–Vung Tau province is facing a number of problems related to wastewater and solid wastes that may cause environmental pollution. The total wastewater volume from seven industrial zones operating in the province is around 18,450 m³/day of which the volume collected and treated is approximately 17,820 m^3/day, accounting for 96.5 %. However, some companies in industrial zones need to build up or upgrade their internal wastewater treatment system before discharging to the environment to minimize water pollution. Moreover, oil and gas enterprises operating in communes of Thang Nhat, Vung Tau province have developed a big industrial zone of nearly 300 ha, but no wastewater treatment system has been provided accordingly. The pollution threats from industrial production and domestic wastewater are significant if there is no wastewater treatment system in place for this newly developed oil and gas industrial zone.

Activities of aquatic farming and processing are also threats to the environment. In reality, there are in total 178 aquatic production units in the province. Among which, there are 32 units for dried products, 31 units for fish sauce, 62 units for aquatic product processing and 38 units for processing of frozen aquatic products for export. Among 178 aquatic products units, there are only 93 that have wastewater treatment system (52 %). And among those 93 units, only 15 meet permissible limits of Vietnam national technical regulations. Hence, pollution from aquatic production activities is becoming serious (Bà Rịa–Vũng Tàu 2012; Nguyễn et al. 2010).

4.5 Impacts of Climate Change

Temperature

According to statistics from 1980–2009, the average annual temperature in the Ba Ria–Vung Tau province increased 0.024 °C/year. The average annual temperature in this period was 27.3 °C.

Two scenarios were used in this study, the medium emission scenario B2 (as recommended by the Ministry of Environment and Natural Resources) and the high emission scenario A1F1.

The computation of scenarios B2 and A1F1 showed that temperature tends to increase every year with the increment of 1.2–1.7 °C by 2050 and 2.3–2.8 °C by 2100 (Figs. 4.2 and 4.3). Future distribution of average temperature does not significantly vary between areas in the provinces. Temperature tends to increase less in the coastal areas to the east and south of the province and increase more in the west and northwest of the province to the mainland (Nguyễn et al. 2010).

Precipitation

According to observations from 1980–2009, precipitation in the Ba Ria monitoring station increased around 3 mm/year. However, in the Vung Tau monitoring station, precipitation decreased around 8 mm/year. The increased precipitation was mainly in rainy season from May to September, accounting for 70 % of the annual precipitation.

Computations from two climate change scenarios showed that the future precipitation tends to increase by 2050 under scenarios B2 and A1FI, by 2.9

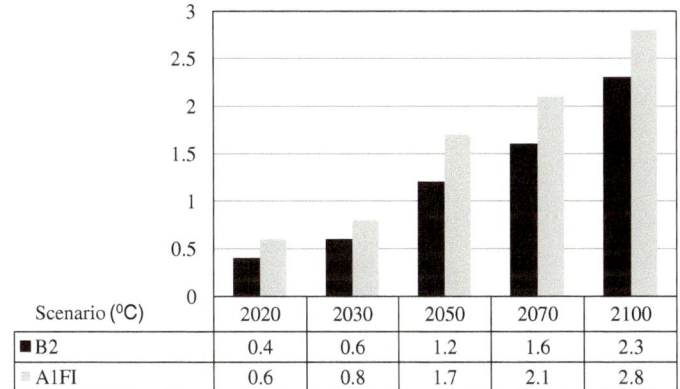

Scenario (°C)	2020	2030	2050	2070	2100
■ B2	0.4	0.6	1.2	1.6	2.3
▨ A1FI	0.6	0.8	1.7	2.1	2.8

Fig. 4.2 Change in annual temperature (°C) compared to baseline (1980–2009) under climate change scenarios B2 and A1FI in the Ba Ria–Vung Tau province from 2020 to 2100

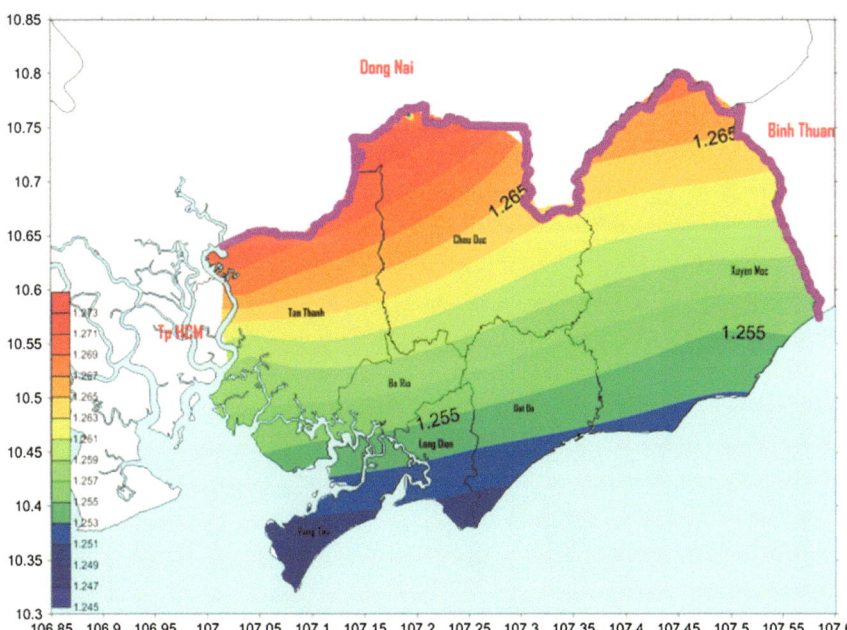

Fig. 4.3 Temperature change (°C) in the Ba Ria–Vung Tau province by 2050 under the B2 scenario compared to baseline (1980–2009)

and 3.5 %, respectively. By 2100, the increase in precipitation is estimated to be 5.5 % (B2) and 8.7 % (A1FI). Regarding the distribution of changes within the province, precipitation increase is not even over the area (Figs. 4.4 and 4.5), (Nguyễn et al. 2010).

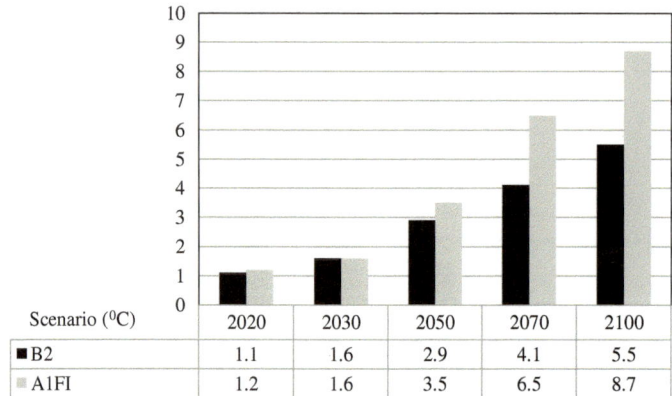

Scenario (°C)	2020	2030	2050	2070	2100
■ B2	1.1	1.6	2.9	4.1	5.5
■ A1FI	1.2	1.6	3.5	6.5	8.7

Fig. 4.4 Change in annual precipitation (%) compared to baseline (1980–2009) under climate change scenarios B2 and A1FI in the Ba Ria–Vung Tau province from 2020 to 2100

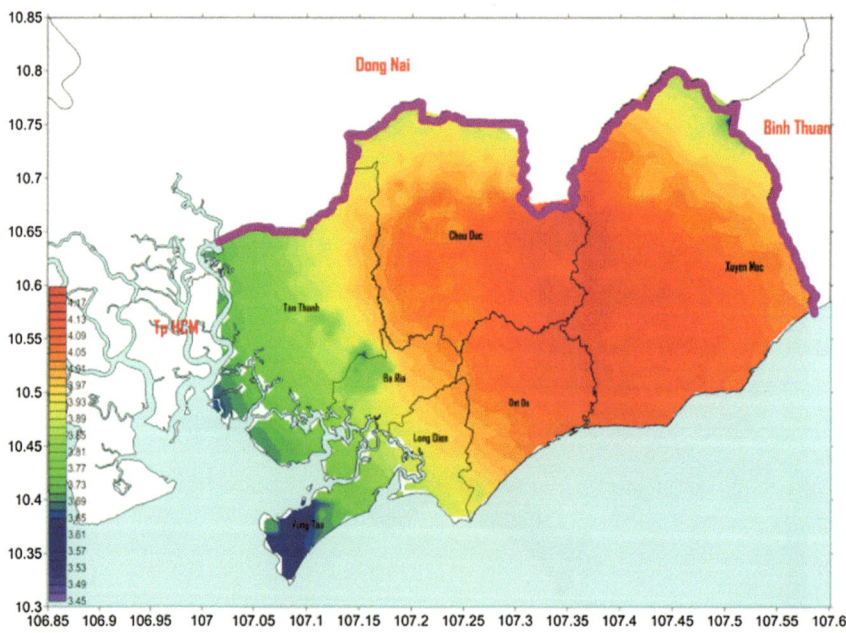

Fig. 4.5 Change in annual precipitation in the Bà Rịa–Vũng Tàu province by 2050 under the B2 scenario compared to the baseline (1980–2009)

Impacts of sea level rise

According to observations at the Vung Tau monitoring station, the sea level in Vung Tau rose by 3 mm/year from 1978 to 2009, with an annual peak sea level around +4 mm and the lowest sea level around −1.5 mm.

The sea level tends to rise from 2020 to 2050 with equivalent increments under scenarios B1, B2 and A1F1. From 2050 to 2100, sea level rise under scenario A1F1 is higher than according to other emission scenarios. The sea level rise by 2050 is projected to be between 26 and 30 cm and that by 2100 between 66 and 99 cm (Fig. 4.6).

Computation of a scenario with one meter sea level rise showed that the flooded area of the Vung Tau province is 5.9 % of the total land area of which the coastal cities are the most vulnerable. Of these, Tan Thanh Districts and Vung Tau City are most seriously influenced having the largest flood prone areas, including industrial zones, riverine and marine ports and protected forest land (mostly mangroves).

Impacts of climate change on surface water resources
The annual average precipitation under different climate change scenarios tends to increase by 2020, 2050 and 2100 as compared to the baseline (1980–1999). Computation shows that the annual average runoff tends to increase accordingly. In Figs. 4.13 and 4.14 the change in annual flow is shown for the rivers Ray, Thi Vai, Du Du and Dinh based on emission scenarios B2 and A1FI. The biggest changes are seen in the flow of the Ray river which increases about 0.04 % by 2020, 0.15 % by 2050 and 0.18 % by 2100. Under scenario A1F1, the flow of the Ray River increases about 0.21 % by 2100 (Figs. 4.7 and 4.8).

Average runoff in rainy season
In the Ray river catchment, the average precipitation during rainy season (as compared to the baseline 1980–1999) by 2020 might increase by 5.4 % and by 2100 by 8 % under the medium emission scenario B2. The increment is even higher in the high scenarios (Figs. 4.9 and 4.10).

Average runoff during the dry season
Average precipitation during dry season varies significantly among rivers. As compared to the baseline (1980–1999), the average precipitation in the dry season by 2100 may decrease more than 3.6 % in the Ray river and more than 3.4 % in

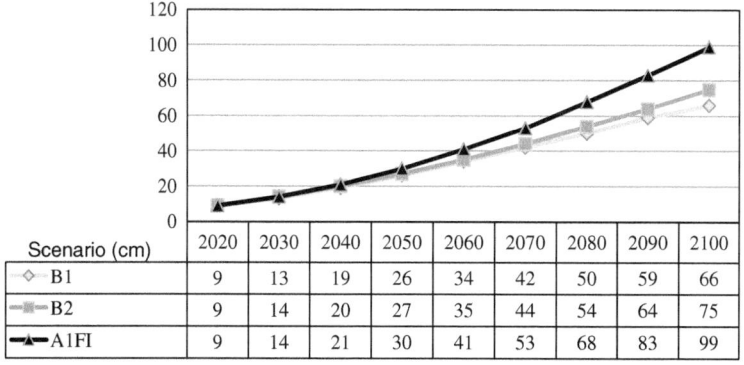

Fig. 4.6 Sea level rise (cm) in the Ba Ria–Vung Tau province under scenarios B1, B2 and A1FI from 2020 to 2100 (MONRE 2012)

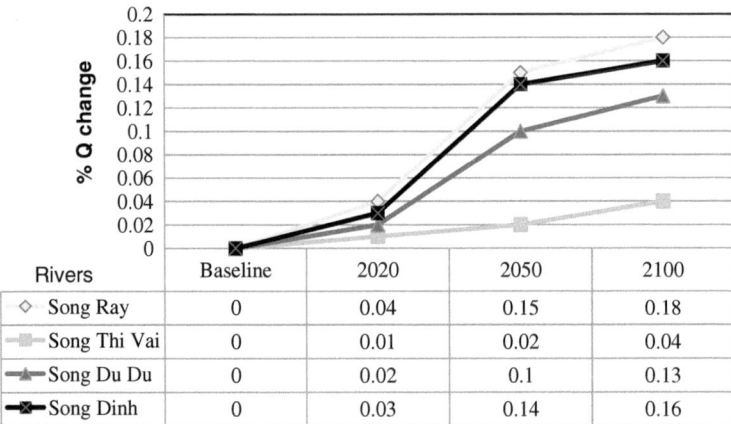

Rivers	Baseline	2020	2050	2100
◇ Song Ray	0	0.04	0.15	0.18
▢ Song Thi Vai	0	0.01	0.02	0.04
▲ Song Du Du	0	0.02	0.1	0.13
▣ Song Dinh	0	0.03	0.14	0.16

Fig. 4.7 Change of annual flow (%) in the rivers of the Ba Ria–Vung Tau province from 2020 to 2100 under scenario B2 as compare to the baseline (1980–2009)

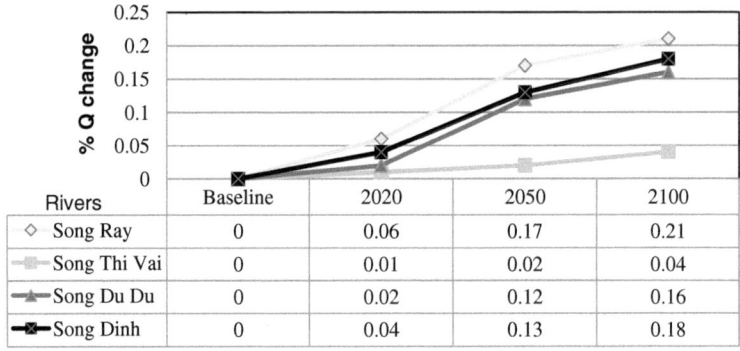

Rivers	Baseline	2020	2050	2100
◇ Song Ray	0	0.06	0.17	0.21
▢ Song Thi Vai	0	0.01	0.02	0.04
▲ Song Du Du	0	0.02	0.12	0.16
▣ Song Dinh	0	0.04	0.13	0.18

Fig. 4.8 Change of annual flow (%) in the rivers of the Ba Ria–Vung Tau province from 2020 to 2100 under scenario A1F1 as compare to the baseline (1980–2009)

the Dinh river according to the average scenario B2. This is a concerning issue that calls for water storage to ensure water supply during dry seasons (Figs. 4.11 and 4.12).

Climate change has impacts on changes of flows in all rivers in the Ba Ria–Vung Tau province. Under climate change scenarios, river flows increase in rainy season due to higher precipitation and decrease during dry seasons.

Impacts of storms and floods

According to statistics from 1962 to 2012, the province has experienced altogether 18 storms and tropical depressions. In the period 1962–2002, the province experienced ten storms, and the last 10 years witnessed eight storms and tropical depressions. Tropical depressions from the East Sea are tending to change their

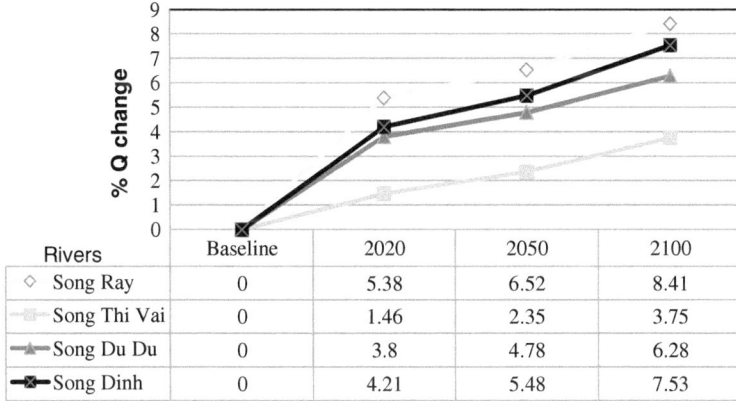

Rivers	Baseline	2020	2050	2100
◇ Song Ray	0	5.38	6.52	8.41
Song Thi Vai	0	1.46	2.35	3.75
Song Du Du	0	3.8	4.78	6.28
Song Dinh	0	4.21	5.48	7.53

Fig. 4.9 Change of annual flow (%) in rainy season in the rivers of the Ba Ria–Vung Tau province from 2020 to 2100 under scenario B2 as compare to the baseline (1980–2009)

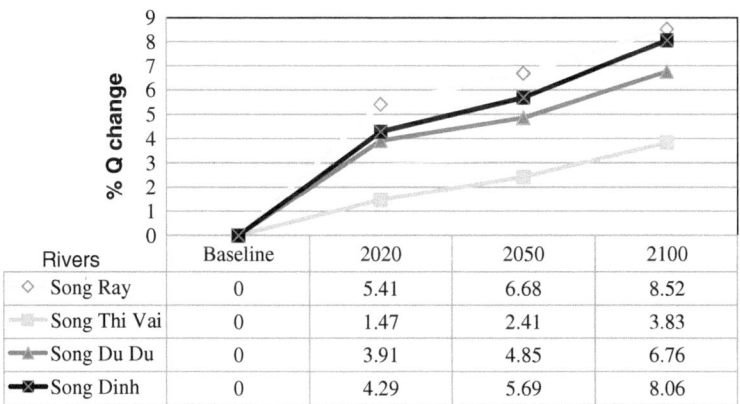

Rivers	Baseline	2020	2050	2100
◇ Song Ray	0	5.41	6.68	8.52
Song Thi Vai	0	1.47	2.41	3.83
Song Du Du	0	3.91	4.85	6.76
Song Dinh	0	4.29	5.69	8.06

Fig. 4.10 Change of annual flow (%) in rainy season in the rivers of the Ba Ria–Vung Ta province from 2020 to 2100 under scenario A1F1 as compare to the baseline (1980–2009)

course to move to the south of Vietnam. Examples of outstanding storms that have caused severe damages in term people and property comprise:

- The death toll of the storm Linda (1997) was 109 with 161 injured and 67 missing. Altogether 568 houses were damaged and many boats sunk. Moreover, in Con Dao, Linda also caused serious damages to ecology as it completely damaged several coral reefs, disordered the nests of sea turtles, and damaged 1/3 of primeval forest.
- The death toll of the storm Durian (2006) was 59 with 783 injured and 8 missing. 224 houses collapsed and 62,908 houses were damaged, 83 boats and junks sunk, 20,000 trees and electricity poles collapsed, and aquatic floating farms suffered huge damages.

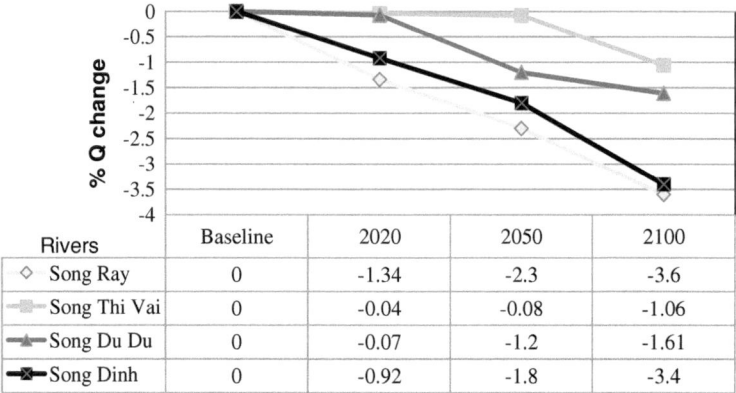

Rivers	Baseline	2020	2050	2100
◇ Song Ray	0	-1.34	-2.3	-3.6
▪ Song Thi Vai	0	-0.04	-0.08	-1.06
▲ Song Du Du	0	-0.07	-1.2	-1.61
✖ Song Dinh	0	-0.92	-1.8	-3.4

Fig. 4.11 Change of annual flow (%) in dry season in the rivers of the Ba Ria–Vung Tau province from 2020 to 2100 under scenario B2 as compare to the baseline (1980–2009)

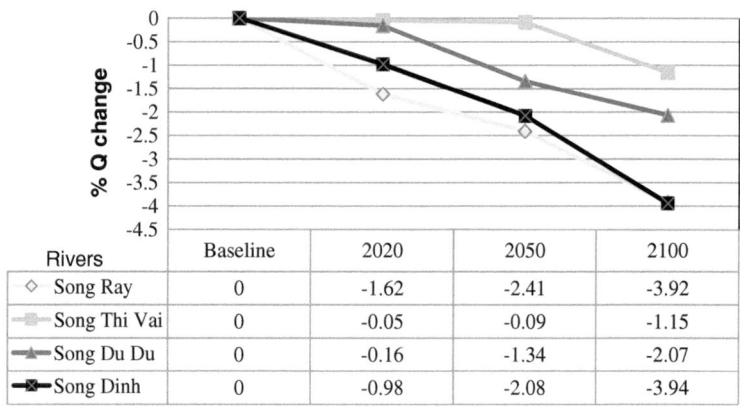

Rivers	Baseline	2020	2050	2100
◇ Song Ray	0	-1.62	-2.41	-3.92
▪ Song Thi Vai	0	-0.05	-0.09	-1.15
▲ Song Du Du	0	-0.16	-1.34	-2.07
✖ Song Dinh	0	-0.98	-2.08	-3.94

Fig. 4.12 Changes of annual flow (%) in dry season in the rivers of the Ba Ria–Vung Ta province from 2020 to 2100 under scenario A1F1 as compare to the baseline (1980–2009)

- The Parkhar storm (2012) was an abnormal storm as it lasted from March 19th–April 1st (dry season) well before the normal storm season (June–November). Having approached the coast of the Ba Ria–Vung Tau province, it weakened to a tropical depression and entered the mainland. The highest wind speed near the storm center was up to level 8. Two persons were injured, 171 houses collapsed, 2,469 houses were damaged, and 47 fishing boats sunk. The damage to salt production was 805 hectares resulting in 15,819 tons of raw salt lost. Altogether 800 trees collapsed and large areas of industrial forest were damaged.

Source: Nguyễn et al. (2010).

Impacts of climate change on estuary erosion and sedimentation

Sediment movement and changes in amount of suspended material in the Cua Lap estuary were simulated under sea level rise scenarios, especifically, under the high emission scenario A1F1 for 2050 and 2100. The simulation results were compared with the baseline in order to assess the impacts of climate change and sea level rise on estuary erosion and sediment accumulation (Figs. 4.13, 4.14 and 4.15).

The computation of suspended sediment concentration samples in the Cua Lap estuary showed that for the baseline scenario, the suspended sediment concentration changes accordingly with higher tides in the estuary. By 2050 and 2100, the sediment concentration will decrease significantly.

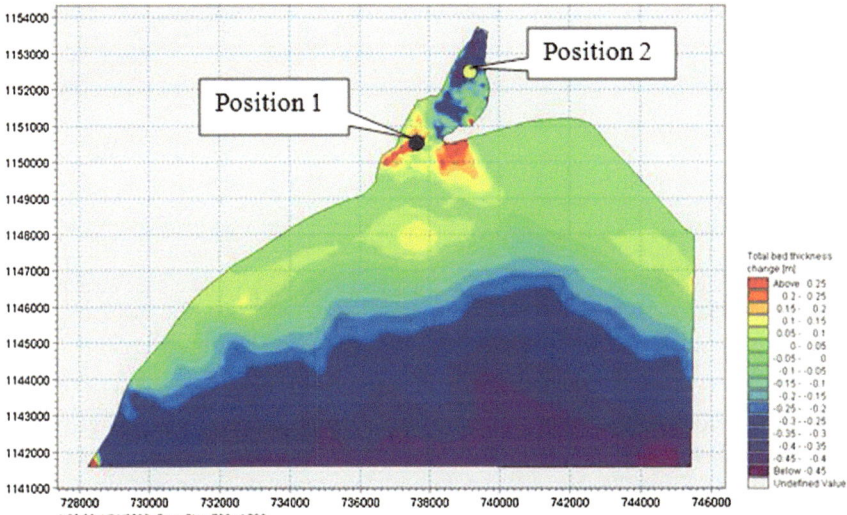

Fig. 4.13 Erosion and accumulation of suspended sediments during one month period in the Cua Lap estuary by 2100

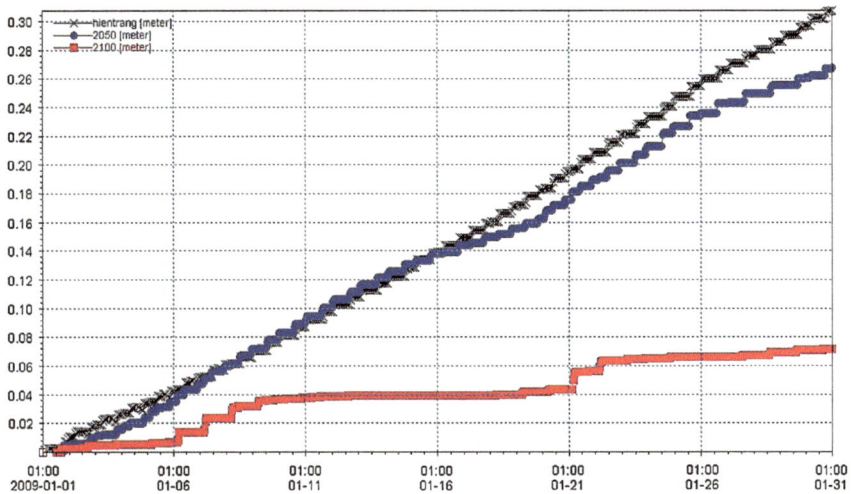

Fig. 4.14 The accumulation of sediments in the Cua Lap estuary (location 1) during one month period for today, in 2050 and 2100 based on the A1FI scenario

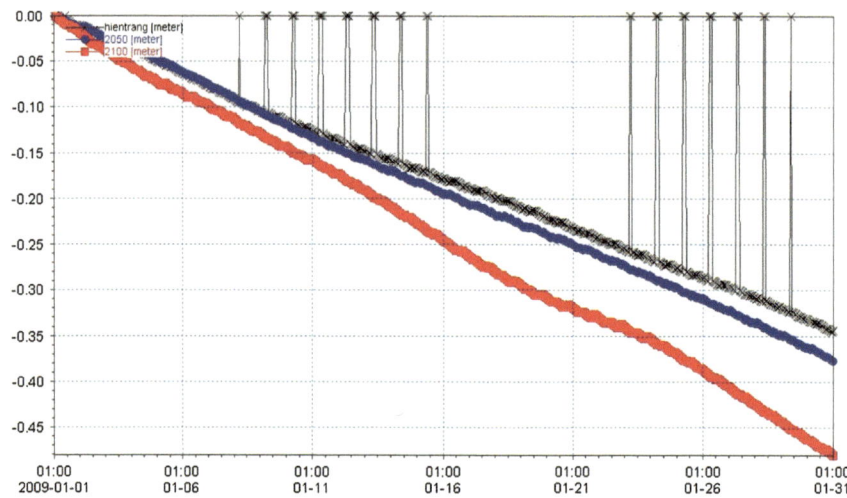

Fig. 4.15 The erosion of sediment in the Cua Lap estuary (location 2) during one month period today, in 2050 and 2100 based on the A1FI

The computation of erosion and accumulation for the Cua Lap area under climate change scenarios shows that in the scenario excluding sea level rise, the estuary is remarkably accumulated. Once sea level rise is taken into account for 2050 and 2100, the accumulation significantly reduced. The result also matches with the computation of decreased suspended sediment in the estuary while the sea level is rising.

These results show that climate change and sea level rise do not increase sediment accumulation at estuaries, but they cause sediment erosion in the inner flow of rivers. Sea level rise definitely has impacts on the erosion and accumulation of sediments at coastal estuaries. Such impacts are quite complicated. Areas that benefit from sea level rise such as estuaries with rather stable terrain can have reduced sediment accumulation which ultimately facilitates the waterway transportation. On the other hand, climate change and sea level rise cause impacts on inner flows of rivers that worsen the erosion, hampering the lives and economic development in the areas.

Impacts of climate change on saline intrusion

The computations and forecasts of climate change show that the Ba Ria–Vung Tau province has been facing and will face problems on natural environment and especially on fresh water resources. Saline intrusion into rivers affects the livelihood and economic development of the province. According to the sea level rise scenarios for the years 2020 and 2030, the increment of the sea level is not significant as compared to the baseline of 2009. But the saline interfaces in the rivers (9 ‰) go beyond the provincial boundaries and reach 10 ‰ at Thi Vai River.

According to the computations, saline intrusion penetrates further into the mainland under sea level rise. Almost all the key rivers are influenced by the saline intrusion.

The economic focus of the Ba Ria–Vung Tau province is not on agriculture, hence the impacts of climate change and saline intrusion has slighter influence on economic development of the province. However, the worsening saline intrusion is causing unexpected consequences for the province and place many difficulties in the future, especially in the provision of fresh water. This trend is strongly affected by increasing exploitation of groundwater leading to a substantial imbalance of the fresh water resources.

4.6 Impacts of Socio-economic Development and Climate Change on Groundwater

Since the Ba Ria–Vung Tau province is rather large, the VIETADAPT project focused on a case study area within the province. The following analyses focuses on socio-economic and climate change impacts for this smaller case study area, taking impacts from the larger province into account, where appropriate.

Population growth and urbanization
The population in the case study area is 106,537 of which urban population accounts for 44.3 %. The average population density is 459 people/km^2 and population working for the agricultural sector accounts for 41.5 %. The annual population growth is 2.7 %, and an increasing urbanization rate raises demands for water supply, and sewage water and domestic waste treatment.

Industrial production development
The case study area is not a heavily industrialized area, but it has important production clusters (Loc An, Bau Xeo, Phuoc Hoi). The overall industrial demand for freshwater is not high, but there is a potential increase in water demand, as well as an increase in the discharge of wastes.

Agriculture production development
Agriculture land area is 112.6 km^2, among which land for rice cultivation is 81.7 km^2 and land for mix cultivation is 30.9 km^2. Irrigation uses mostly surface water, but groundwater is also in use in some places.

Service development
There are five major tourism areas in the case study area, Long Hai, Ninh Dam mountain, Phuong Hai, Ho Tram and Ho Coc. The total demand for water in these areas is 2,000–4,000 m^3/day, of which only a small portion supplied by groundwater (Cục thống kê Bà Rịa–Vũng Tàu 2010).

Water demand
The total water demand of the case study area in 2010 was 47,850 m^3/day of which domestic water demand was 17,240 m^3/day and industrial water demand was 30,610 m^3/day. The main water supply is surface water, and groundwater is partly used via centralized water supply network and local groundwater exploitation. The projected increase in water demand (Table 4.1) suggests that groundwater will become increasingly important for water supply.

Table 4.1 Water demand by districts (m³/day) in 2010, 2105 and 2020

No.	District, town, city	2010		2015		2020	
		Domestic use	Industrial	Domestic use	Industrial	Domestic use	Industrial
1	Long Dien	6,574	20,215	7,203	22,913	13,954	27,580
2	Dat Do	3,423	6,500	3,807	18,869	7,486	33,640
3	Xuyen Moc	7,238	3,900	7,931	5,200	15,363	7,000
Total		17,235	30,610	18,941	41,787	36,803	68,220
		47,845		60,728		105,023	

Source Report on groundwater resource planning in Ba Ria–Vung Tau by 2020

4.7 Impacts of Socio-economic Development on Groundwater Resources

Groundwater resources exploitation

Total groundwater exploitation in the case study area (Fig. 4.16) is 10,427 m³/day, among which domestic water is 7,043.8 m³/day and production water is 3,383.3 m³/day. Details of the groundwater exploitation are presented in Table 4.2. The figures show that the increasing water demand has boosted the exploitation of groundwater, which will pose challenges for future groundwater supply (Ủy ban nhân dân Bà Rịa–Vũng Tàu 2012).

Industrial production and home craft industry

Industrial production activities in the case study area is low, but the industrial production in the surrounding areas have a water use demand of 30,015 m³/day, which is projected to increase to 68,220 m³/day by 2020. Currently, the main water source is surface water. These industrial activities discharge a significant

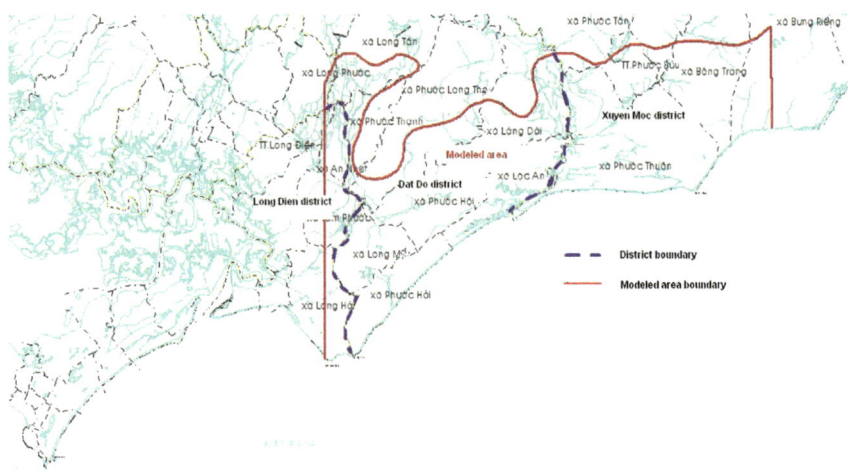

Fig. 4.16 Groundwater modelling case study in the Ba Ria–Vung Tau case study

Table 4.2 Current status of the groundwater exploitation in the case study area

No.	Commune, ward, town	Number of households	Number of wells			Density of wells		Exploitation output (m³/day)		
			Total	Dug wells	Drilled wells	Number of wells (km²)	Number of wells/household	Total	Domestic	Industrial, agricultural, service sector
I	*Long Dien district*									
1	Long Phuoc	73	43	26	17	2.64	0.02	103.8	55.7	48.1
2	No. Long Dien	72	15	3	12	1.03	0.00	16.8	16.3	0.5
3	Tam Phuoc	769	340	94	246	24.73	0.22	349.0	145.0	204.0
4	An Nhut	832	362	162	200	61.17	0.37	350.2	407.1	−56.9
5	Long Hai	4,494	88	3	85	8.01	0.01	73.0	56.0	17.0
II	*Dat Do district*									
6	Phuoc Hoi	1,386	644	464	180	28.36	0.46	1,884.0	644.0	1,240.0
7	Loc An	912	124	117	7	7.00	0.14	235.0	124.0	111.0
8	No. Phuoc Hai	5,393	1.936	1.443	493	123.55	0.36	1,406.0	1,355.0	51.0
9	Long My	978	836	748	88	64.36	0.85	1,192.0	836.0	356.0
10	Phuoc Thanh	2,317.95	515.25	502.65	12.60	23.27	0.10	494.05	361.50	132.6
11	Phuoc Long Tho	362.92	355.18	319.92	35.26	9.84	0.42	405.58	115.62	290.0
12	Long Tan	648.00	403.65	220.95	182.70	13.72	0.28	304.85	261.70	43.2
13	Lang Dai	976.95	640.90	336.70	304.20	27.53	0.43	442.30	333.10	109.2
III	*Xuyen Moc district*									
14	No. Phuoc Buu	2.806	1.582	470	111	172.00	0.45	4088	382.4	26.4
15	Phuoc Tan	334	316	33	283	9.77	0.09	504.9	315.8	189.1
16	Bung Rieng	724	623	337	287	12.47	0.52	423.3	394.2	29.1
17	Bong Trang	630	509	42	467	14.60	0.61	485.3	375.8	109.5
18	Phuoc Thuan	1,959	1.026	191	835	20.26	0.52	899.0	621.0	278.0
19	Xuyen Moc	967	552	50	502	30.36	0.20	449.2	243.7	205.5
	Total	**26.634**	**10.910**	**5.563**	**4.346**	**41**	**0.41**	**10,427.0**	**7,043.8**	**3,383.2**

Source Report on groundwater resource planning in Ba Ria–Vung Tau by 2020

amount of wastewater and solid waste that cause environment pollution, affecting also the groundwater resources. This is challenging not only because of the
state of the environment, but also because groundwater resources are projected to
become more important in the coming decades.

Agricultural production activities

Agriculture is economically not very important in the Ba Ria–Vung Tau province.
Nevertheless, apart from groundwater exploitation, agricultural activities pose
threats to water quality due to the use of pesticides and fertilizers. So far, studies did not find high pesticide residues in groundwater, except for nitrogen compounds. Currently, surface water is mostly used for irrigation and groundwater
plays a minor role, which might change in the future.

Tourism service activities

Tourism is economically very important for the Ba Ria–Vung Tau province
in general and the case study area in particular. According to statistics from the
Department for Culture, Sport and Tourism, the Ba Ria–Vung Tau province
receive more than 8 millions tourist annually, among which 300,000 are foreign-
ers. The total water demand for tourism is 16,000 m^3/day, of which a small portion is groundwater. Traditionally Vung Tau city was the hub for tourism of the
province, but recently tourism has spread beyond the city. Despite the economical
benefit for the province, tourism also poses a potential risk for environment pollution and an increase to the water demand, including groundwater (Bà Rịa–Vũng
Tàu 2010b, 2012).

Socio-economic development activities in the province pose potential threats
to the entire groundwater system, including: (i) over exploitation of groundwater
due to uncontrolled groundwater use; (ii) pollution of surface water due to waste
dumping to channels and canals, which ultimately pollutes groundwater resources
through seepage of contaminants; and (iii) saline intrusion due to groundwater
overexploitation of coastal aquifers.

Increase of natural recharge of groundwater during rainy seasons

Results of model calculations show that climate change enhances the components
of natural recharge, except the recharge component for rivers. This leads to an
increased water level in aquifers and could be a positive sign since it increases the
volume of natural groundwater resources (Table 4.3).

Saline intrusion

It is notable that the groundwater flow from the sea (saline water) is increasing,
even though to at small amount (157 m^3/day—rainy season 2100). This shows that
the process of saline intrusion will increase locally in some coastal aquifers. Saline
intrusion under scenario A1F1 is stimulated and forecasted using the SEAWAT
model for dry season (Table 4.4 and Fig. 4.17).

Computation from SEAWAT model shows that two topmost aquifers experience the highest increase of saline intrusion under scenario A1F1.

Table 4.3 Main components of natural recharge of aquifers in the case study area

Elements	Amount of water (m³/day)										
	2012	2020	2030	2040	2050	2060	2070	2080	2090	2100
River	9,410	9,452	9,413	9,396	9,308	9,251	9,133	9,076	9,020	8,917
Recharge by precipitation	10,992	11,448	11,515	11,515	11,734	11,734	12,120	12,120	12,120	12,398
Infiltration from upper aquifer	18,867	18,944	18,892	18,908	18,970	18,958	19,067	19,052	19,046	19,130
Infiltration from the sea	931	960	987	1,029	1,078	1,139	1,211	1,315	1,431	1,572

Table 4.4 Estimations of areas affecting saline intrusion in different aquifers within the case study area

No.	Aquifer	Areas of saline intrusion distribution by years (m²)						
		2012	2020		2050		2100	
		Area	Area	Increase in comparison to 2012	Area	Increase in comparison to 2012	Area	Increase in comparison to 2012
1	The semipermeable layer Q₂	13,804,600	13,832,209	27,609	14,011,669	207,069	21,742,100	7,937,500
2	qh layer	30,010,000	30,136,042	126,042	30,310,100	300,100	39,385,000	9,375,000
3	qp layer	23,540,000	23,617,682	77,682	23,634,160	94,160	23,681,240	141,240
4	Layer n₂²	11,130,000	11,174,520	44,520	11,177,859	47,859	11,191,215	61,215
5	ms layer	16,810,000	16,877,240	67,240	16,880,602	70,602	16,880,602	70,602

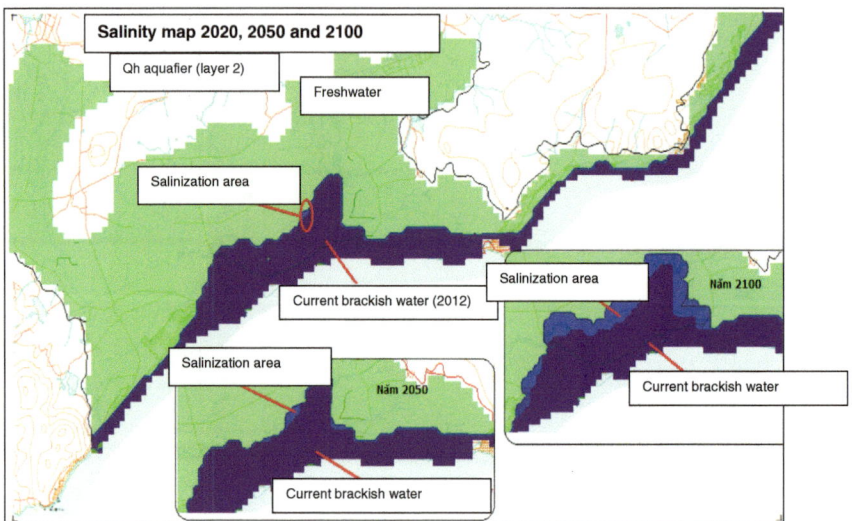

Fig. 4.17 Saline intrusion simulation of the semipermeable aquifer (Q2) (layer 1) in the case study area

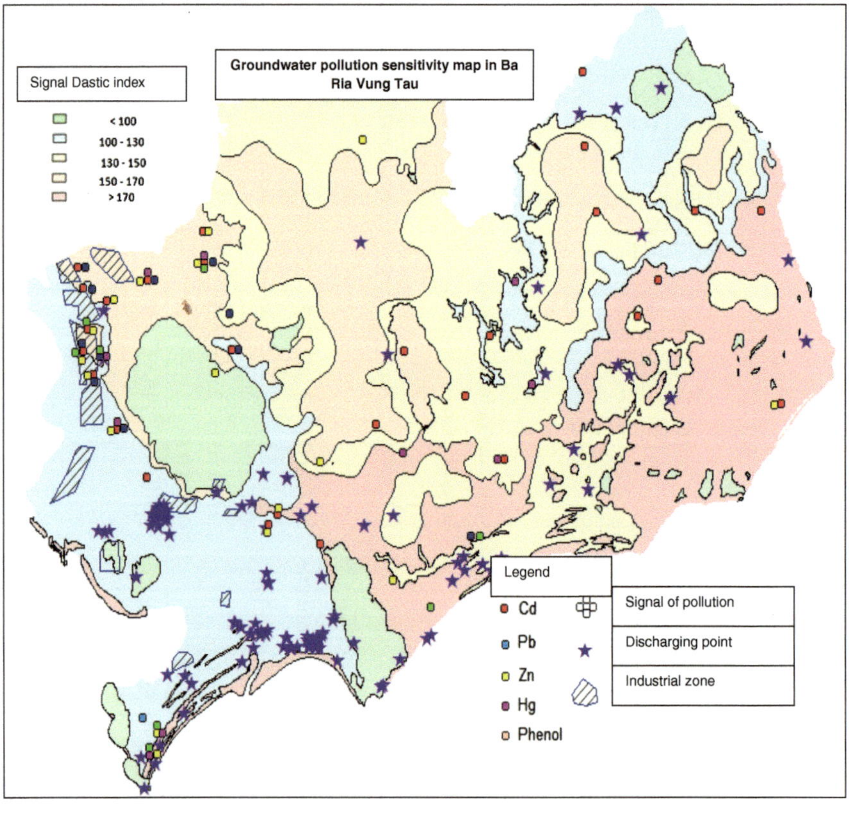

Fig. 4.18 Groundwater pollution risk map of the case study area based on the DRASTIC evaluation

Table 4.5 Volume of the water that is infiltrated into aquifer for 2012–2100 in the case study area. Estimations are based on A1FI scenario

Season	Vertical infiltrated water amount (m³/day)									
	2012	2020	2030	2040	2050	2060	2070	2080	2090	2100
Dry season	16,063	16,063	16,040	16,046	16,031	16,015	15,996	15,977	15,969	15,960
Rainy season	18,867	18,944	18,892	18,908	18,970	18,958	19,067	19,052	19,046	19,130

Increasing risk of groundwater pollution by waste deposition on the land and in the rivers

Socio-economic development activities discharge wastewater and wastes into the land and water bodies. The case study area has the DRASTIC index (U.S. EPA 1987) of over 170, which turns it into a groundwater pollution sensitive area (Fig. 4.18). This means that as soon as there is no solution for dealing with potential harmful elements on the ground, the danger of groundwater pollution is extremely high. In addition, computation of flow models for groundwater recharge also showed that the high soil permeability tends to increase significantly groundwater recharge in rainy seasons, which will increase the seepage of contaminants. Hence the risk of groundwater pollution from wastes deposited on permeable soil poses a serious challenge. Groundwater recharge might reduce slightly during dry seasons (Bà Rịa–Vũng Tàu 2010a) (Table 4.5).

References

Bà Rịa–Vũng Tàu (2010a) Đề án xử lý ô nhiễm tỉnh Bà Rịa–Vũng Tàu giai đoạn 2011–2015, UBND tỉnh Bà Rịa–Vũng Tàu (Ba Ria–Vung Tau (2010) Scheme on polluted water treatment in Ba Ria–Vung Tau period 2011 to 2015, Ba Ria–Vung Tau)

Bà Rịa–Vũng Tàu (2010b) Điều chỉnh Quy hoạch tổng thể phát triển kinh tế—xã hội thị xã Bà Rịa–Vũng Tàu thời kỳ đến năm 2020, Bà Rịa–Vũng Tàu (Ba Ria–Vung Tau (2010) Adjustment of the master plan of social economic development of Ba Ria–Vung Tau by 2020, Ba Ria–Vung Tau)

Bà Rịa–Vũng Tàu (2012) Kết quả thực hiện nhiệm vụ phát triển kinh tế - xã hội giai đoạn 2006–2010 và kế hoạch phát triển kinh tế—xã hội giai đoạn 2011-2015 tỉnh Bà Rịa–Vũng Tàu (Ba Ria–Vung Tau (2012) Results of the implementation tasks of economic development—social plan for 2006–2010 and economic development—society in 2011–2015 Ba Ria–Vung Tau)

Cục thống kê Bà Rịa–Vũng Tàu (2010) Niên giám thống kê tỉnh Bà Rịa–Vũng Tàu. Bà Rịa–Vũng Tàu (Ba Ria Vung Tau statistical office (2010) Ba Ria–Vung Tau's statistical yearbook, Ba Ria–Vung Tau)

U.S. EPA (1987) DRASTIC: a standardized system for evaluating groundwater pollution potential using hydrogeological settings, EPA/600/2-87/035. http://www.epa.gov/region5/waste/clint onlandfill/PDFClintonLFChemicalWaste_USEPAApplication/cl_130.pdf

Nguyễn ĐH, Đặng ĐL,Trần Thị XT (2010) Tác động của biến đổi khí hậu đối với Bà Rịa–Vũng Tàu—Trung tâm Phát triển Xã hội và Môi trường vùng CERSED. Bà Rịa–Vũng Tàu (Nguyen DH, Dang DL, Tran Thi XT (2010) The impact of climate change for the Ba Ria–Vung Tau—Centre for Social Development and the Environment CERSED, Ba Ria–Vung Tau)

MONRE (Bộ Tài Nguyên Môi Trường) (2012) Kịch bản Biến đổi khí hậu, nước biển dâng cho Việt Nam. Hà Nội (Ministry of Natural Resources and Environment (2012) Climate change scenarios and sea level rise for Vietnam. Hanoi)

Ủy ban nhân dân Bà Rịa–Vũng Tàu (2012) Xây dựng kế hoạch hành động ứng phó với biến đổi khí hậu tỉnh Bà Rịa–Vũng Tàu trong khuôn khổ chương trình mục tiêu quốc gia, Bà Rịa–Vũng Tàu (People's Committee of Ba Ria–Vung Tau (2012) Building an action plan to respond to climate change in Ba Ria–Vung Tau in the framework of national objectives, Ba Ria Vung Tau)

Chapter 5
Scenario Workshops

Abstract Scenario workshops are a participatory tool to develop story lines on potential future developments. The scenarios are based on scientific research of the living environment conditions, climate change impacts and socio-economic development impacts. Based on the stakeholder input the scenarios and storylines were continuously improved. The most realistic scenarios were chosen to develop first drafts of climate change adaptation measures. The methodological approach can be used also for other areas in Vietnam and internationally. The scientists involved were able to build up deep trust among the stakeholders and the Ministry of Natural Resources and the Environment has expressed that the scenario workshop methodology shall be implemented also in other areas in Vietnam to develop climate change adaptation measures.

Introduction

The overall target of the VIETADAPT project was to contribute to the climate change adaptation strategy of Vietnam via the development of action plans at provincial level and adaptation solutions. The project developed pilot models for climate change adaptation that could also be implemented in other provinces in Vietnam.

The provincial-level climate change adaptation studies in Vietnam has created noticeable signals to stakeholders such as functional authorities (Women's Union, Committees for flood and storm control and others) and provincial authorities and departments to react. The response to climate change must be implemented by a combination of downscaled climate change models and impacts studies, considering trends and plans for socio-economic development at provinces. Current and future vulnerability patterns play a crucial role in the identification of adaptation options. Hence, to develop feasible adaptation solutions for provinces, it is necessary to organize workshops that support close cooperation between scientists, stakeholders and local authorities.

Scenario workshops are a participatory approach tool that supports communication processes among stakeholders, scientists and decision makers. The main strength of scenario workshops is the development of long-term storylines of possible future developments. These storylines are a suitable tool to integrate natural

P. Schmidt-Thomé et al., *Climate Change Adaptation Measures in Vietnam*,
SpringerBriefs in Earth Sciences, DOI 10.1007/978-3-319-12346-2_5

hazards and climate change impacts into land-use planning. The VIETADAPT project followed the methodologies and experiences of the BaltCICA project. The scenario workshop methodology applied in the BaltCICA project was elaborated by the Danish Board of Technology and implemented in the city of Kalundborg (Bedstedt and Gram 2013). Based on these experiences the methodology was developed further to address the respective requirements of other case studies in the Baltic Sea Region (Schmidt-Thomé and Klein 2013).

Background of scenario workshops

Participatory approaches have proven to be a most valuable tool in decision-making, not only because governance issues and the involvement of citizens in decision-making practices are becoming more important from a basic democratic and social justice perspective. Participatory approaches have proven to be a rather efficient tool because the direct involvement of various interest groups not only leads to a more complex understanding of the issues at stake but also to developing socially accepted solutions to problems, taking into account various interests and expertise. Applying interdisciplinary approaches in decision making processes might appear to be challenging but it has turned out that the early integration of manifold expertise and interests minimizes the risk of costly adjustments at a later stage (Slocum 2003; Wollenberg et al. 2000). In addition, the integration of relevant stakeholders at an early phase reduces the amount of potential resistance and therefore leads to an overall quicker implementation of measures (e.g. Rimkus et al. 2013; Petersell et al. 2013).

There are several different participatory approaches. The "scenario workshop" was chosen for the VIETADAPT project as it is one tool that specifically addresses potential but unknown complex future developments and a long time horizon. According to Slocum (2003) scenario workshops are useful to, e.g. improve overall preparedness and long-term decision making as well as to develop alternative options for future developments. These characteristics are of particular importance when dealing with land-use planning.

Scenarios are storylines of possible future changes, not predictions or projections. Such storylines are used to identify potential future developments and to react timely by taking early measures. Here, the aim was to safeguard economic development and social safety by early decision making on appropriate adaptation measures.

Preparation of scenario workshops

Slocum (2003) has written a very concise overview on several participatory approaches containing very detailed guidelines on how to organize workshops, select and invite participants and other important issues. This paper does not intend to rewrite these existing guidelines. But there are some issues on the particular topic of implementing natural hazard and climate change adaptation that are discussed in this chapter.

Geological processes happen in rather long time-scales, at least from a human perspective. Several geohazards and related georisks may never occur in a human life, others might occur rather frequently. Also anthropogenic contamination of the environment is not necessarily visible within a human life span of, e.g. 80–100 years, and might in fact affect the environment long after contaminating human activities have ceased. The length of time spans, geological processes and impact delays are therefore one very important issue in the communication of geosciences. As one example of the many in which scenario workshops might be applied, climate change is only one example. During the BaltCICA project it became obvious that climate change is certainly an ongoing process but usually people tend to focus on those climate change impacts that might be expected after approximately 100 years. This is mainly due to the reporting horizon of most models and scenarios presented in the Intergovernmental Panel on Climate Change's (IPCC) reports. But it is often forgotten that climate change is an ongoing, slow process which impacts occur slowly. Also, all used models and scenarios have large uncertainties. It is certainly possible that those rather adverse impacts occur, but it is also possible that these do not occur, or that the impacts are differently than currently perceived. It is therefore very important to keep the timescale of investments and/or land use changes in mind. For example, some investments that yield revenues in a short term horizon are possibly not even affected by climate change, or might be decommissioned in case of adverse climate change effects. In the case of adaptive measures, it is vital to analyze if planned investments are beneficial already today or only in the case of possible climate change impacts.

It is thus one important part of the workshop preparation to discuss the time horizon of measures versus the timescale of geological and environmental processes. This might strongly affect the stakeholder's points of view and thus influence the decision making.

Ideally, all possibly involved or affected stakeholders should be invited to scenario workshops, but this is not always possible for practical reasons. It should therefore be taken good care of inviting representatives of the most important stakeholder groups of a particular issue at stake. Also, it is of great importance to invite stakeholders that have a potential to obstruct investments. By involving also critical persons it is possible to avoid resistance at a later stage.

It is essential to prepare sufficient but concise workshop material at an early stage and provide this to all workshop participants for preparation. If possible, feasibility studies and/or cost estimations of potential investments shall also be made available. The workshop time is usually rather limited and the time spent for general understanding should be kept as short as possible. The workshop is most effective when participants focus mainly on the matters to be discussed and are not distracted by other explanations. A briefing meeting prior to the workshop makes sense in case the participants are not familiar with the issue. Such a meeting is also most valuable as the participants can present their views of the problem and their expectations towards solutions.

Early definition of workshop aims
It is recommendable to pre-formulate possible workshop outcomes; i.e. what kind of decisions shall be taken at the end of the workshop? It is therefore

recommendable to plan for at least two workshop sessions, as experience has shown that many people need the first workshop to fully understand the topic and the expectations behind the workshop. Usually, the workshop group is better informed and more focused during the second workshop session so that decision making goes smoother.

5.1 Scenario Workshops Applied in the VIETADAPT Project

The workshops were prepared long-term in advance by establishing contacts with local stakeholders. Once stakeholders confirmed their interest to participate, appropriate venues were identified, and invitation letters containing agendas were sent out. All workshops started with the official registration of participants and a group photo for official documentation.

Climate change might lead to changes in extreme natural events, but does not necessarily do so. In fact, proven impacts of climate change on extreme events are rare and are usually characterized by substantial uncertainties (IPCC 2012). On the other hand, it is indisputable that costs and the number of affected people by natural hazards are constantly rising. There is no doubt that this development is due to changes in vulnerability patterns (IPCC 2012). In other words, the increasing risk related to natural hazards is definitely attributable to changes in the exposure of humans and assets, by direct activities (e.g. building in flood prone areas), indirect impacts (deforestation leading to landslides) or mal-adapted land use (intensive agriculture in dry areas).

Unfortunately, climate change is often quickly used to explain the rising costs and other impacts of natural hazards. It appears that climate change is thereby used as a scapegoat to divert from the main underlying causes, i.e. vulnerability patterns. But instead of blaming vague factors, such as a changing climate, the fact is that human vulnerabilities should be carefully assessed. Only if current vulnerabilities and resulting risk patterns are understood appropriate solutions can be developed. And only then it is possible to feasibly discuss and probably respect potential climate change impacts and implement appropriate climate change adaptation measures in the decision making process.

The VIETADAPT project has primarily focused on identifying both current as well as potential future vulnerability patterns in its case study areas including identification of appropriate climate change adaptation measures. High priority was given to distinguishing between various vulnerabilities (physical, social, economical and environmental). This distinction is very important because vulnerabilities affect an area differently and may be tackled by various institutions. On the other hand, a holistic approach distinguishing between the vulnerabilities is necessary to understand potential interactions and thus sensibilize stakeholders towards interdisciplinary approaches to mitigate the overall vulnerability.

The project started with collecting existing datasets from different institutions and organizations, mainly from local Departments of Natural Resources and Environment (DONRE), in order to have an overview of the current environmental and socio-economical status of the case study areas. In addition to use existing datasets some field work studies were carried out, especially to examine groundwater conditions. Data was also collected via interviews and questionnaires.

Implementing the scenario workshops

The VIETADAPT project conducted two Sharing, Learning, Dialogue (SLD) workshops in each case study. The intention was to build up trust, gain experience and inputs from local stakeholders. The first workshop was held in April, 2012 in the two case studies Thanh Hoa and Ba Ria–Vung Tau provinces. In an intermittent project meeting in February 2013, preliminary results were assessed and discussed among the project team. In the following, the case studies were visited to discuss the progress also with the local stakeholders. The second round of SLD workshops were held in May, 2013. During these workshops the project team presented modeling and water balances results and asked for direct feedback on first ideas on climate change adaptation measures. The final workshop that presented the preliminary adaptation measures was held in November 2013.

The first round of workshops

- *Venue*: The Guest House of the Ministry of Natural Resources and Environment, Sam Son Town, Thanh Hoa Province and Post Hotel, Vung Tau Province.
- *Time*: April 18th 2012 and April 25 2012
- *Participants*: 25 participants (for the workshop in Thanh Hoa province) and 32 participants (for the workshop in Ba Ria–Vung Tau province), including project members from the The Geological Survey of Finland, the Vietnam Institute of Meteorology Hydrology and Environment and the National Center for Water resource planning and Investigation, local Departments (DONRE, local centers for Hydro-Meteorological Forecasting etc.) and local media (Figs. 5.1 and 5.2).

Normally, participants are provided with workshop scenarios before coming to the workshop (Bedstedt and Gram 2013). However, this approach was not applied in the VIETADAP project. Discussions with local agencies showed that such scenarios may mislead participants and thus make discussions with stakeholders ineffective. Instead, the project team started the workshop by presenting the status quo of the environment and economic situation of the target province, and potential impacts of climate change. Participants (local) were then requested to discuss future development plan and build up their own scenarios.

Since this approach was new to both project team and stakeholders in Vietnam, the first workshop was a chance to test the ability and desires of local stakeholders via discussion on future development and potential climate change impacts. This was important to test participant's understanding of vulnerability concept and

Fig. 5.1 Discussion sessions in the workshop at Thanh Hoa and Ba Ria–Vung Tau provinces

Fig. 5.2 Interview with local newspapers and group photo of participants

how the discussion could be further developed. The purpose of the first workshop was to improve the comprehensive approach as well as to indicate the necessity of cooperation between data owners in the project.

The first workshop showed that stakeholders master an extensive knowledge of socio-economic structure and environment in their province. The importance to integrate environmental impacts in land-use planning is repeatedly mentioned in the discussions. The vulnerability concept was new to almost local participants. Yet once it was grasped, the discussion quickly developed. The discussion also covered the contact and approach to the new data set between project team and local stakeholders Final agreement was that the exchange of data shall facilitate the project and hence the latter would receive better data from different stakeholders.

The first workshop was a chance for the project team to learn experience for further workshops. Participants also made contribution to the project by indicating the most crucial demands of their province that need researches regarding natural disaster, zones of vulnerability and climate change adaptation measures. Participants also proposed that project should focus more on specific areas in target provinces, which means the real hot spot. The reason was that target provinces of the study were large, hence it was difficult to be fully covered in one workshop. Moreover, once the workshop proves success at compact scale the result could be easily extrapolated for developing the 10 master plan.

The second round of workshops

- *Venue*: Hoa Lu Hotel, Ninh Binh city and Muong Thanh Hotel, Vũng Tàu city.
- *Dates*: 08/5/2013 and 15/5/2013
- *Participants*: 42 attendees including: the Geological Survey of Finland (GTK), the Vietnam Institute of Meteorology, Hydrology and Environment (IMHEN), the National Centre for Water Resources Planning and Investigation (CWRPI), local agencies (The People's Committee, Department of Natural Resources and Environment, Science and Technology, Agriculture and Rural Development, Transport, Construction, Trade and Industry, Planning and Investment, Culture-Tourism, Center for Hydro-meteorology forecasting, Natural disaster prevention committee, Women Union…) and members of the press.

The second workshops were held one year after the first ones. In order to prepare the stakeholders for holistic discussions on the vulnerability patterns and climate change adaptation measures, the project developed a questionnaire that was sent out to all participating stakeholders prior to the workshop. The workshop discussions among stakeholders and project team were supported by giving presentations about temperature, precipitation and sea level rise scenarios for 2030, 2050 and 2100 in the case study area, based on two different Forcing Scenarios (B2, A1F1). Current and extrapolated future water balance figures were presented for groundwater and surface water resources, including estimations on saline intrusion to aquifers and rivers. The future scenarios of water balance took changes in water use within different sectors (private, industrial, irrigation) and within different climate conditions (water discharge and recharge) into account. Further socio-economic and environmental development figures within case study area were presented. Finally, some climate change adaptation measures were introduced as a basis for the discussions.

The guided discussion was organized by using the questionnaires (see Annex) as a stimulator. The questionnaires also served as a guideline to concentrate the discussions on the vulnerability patterns before entering the analysis of climate change adaptation options. Those stakeholders that had not filled in the forms beforehand were asked to do it during the guided discussion. It was also possible to fill in the forms by those stakeholders that had done it already, especially in case of misunderstandings of technical terms. The guided discussion within the workshop gave the possibility for the project team to explain more of the underlying scientific concepts in detail, meanwhile the stakeholders had the opportunities to ask questions and give direct feedback to the project.

The questionnaires (Questionnaires form in the Annex) that were filled during the workshops yielded different results than those filled in before. It became obvious that many stakeholders had rather underestimated the impacts of climate change and socio-economic developments on groundwater resources as well as the overall drinking water supply. It was therefore interesting to observe that water issues gained rather strongly importance once the stakeholders were sensibilized towards them.

Also other questionnaire results slightly changed when the questionnaires were filled in during the workshops. This can be explained by the fact that most stakeholders are not trained in geo-sciences. This underlines the importance of communication with stakeholders when discussing climate change adaptation measures. Meanwhile one could argue that such workshops lead to biased results in questionnaires it should rather be seen the other way round: Stakeholders need to be both informed about and sensibilized to climate change and socio-economic impacts on the living environment before drafting adaptation methods.

Initial outcomes of the workshops
The workshop concept that had been applied in the VIETADAPT project revealed good results. The spontaneous and direct feedback from stakeholders during the workshops was positive with constructive criticisms and suggestions for future work. The questionnaire was found to be feasible tool for collecting non-existing, empirical data of stakeholders' understanding on natural hazards, vulnerabilities and climate change adaptation measures. The possibility to give own comments additionally to the questionnaires was also received very well which was proven by the great number of free comments in the documents.

The workshops together with questionnaires gave a good basis for developing the first climate change adaptation measures on an interdisciplinary level. The strong involvement of local stakeholders for the case study areas was the key issue.

The first results achieved comprise the acknowledgement of the stakeholders that climate change is ongoing. Further it is now understood that current risks related to natural hazard are affected by climate change, partly because shifts in climatic variables (dry spells/wet periods), and partly because of secondary effects, such as sea level rise. But most importantly, not only climate change affects hazard patterns, the stronger impacts results from human vulnerabilities. For example, higher population densities lead to higher groundwater consumption, which in turn leads to a higher risk of groundwater salinisation. Changes of river courses and deforestation lead to changing flood patterns, and these effects are stronger than the one's of climate change. Also settlements in flood prone areas lead to a higher amount of floods and more people affected than during earlier flood events.

It was acknowledged by the local stakeholders that it is important to adapt to climate change, but that it is equally important to include hazard patterns and socio-economic development trends into account, too. Only by balancing these three main components it will be possible to achieve sustainability in the adaptation process.

In order to finally implement local climate change adaptation options several further steps are important: A still stronger cooperation with a wider variety of the local stakeholders. The stakeholders also expressed their interest in cost-benefit analysis and uncertainty estimations for concrete adaptation measures, especially because the competitiveness of the case study areas must be guaranteed.

Representatives of the local press were invited to all workshops for dissemination purposes. The workshops were then broadcasted in the local news in order to inform the general public upon the project and its contents.

Developing adaptation measures

After the second workshop of the VIETADAPT project, the project team further analyzed the questionnaire and discussion takeaways with focus on feasible climate change adaptation measures that was drawn from the workshop. For the first time, climate change adaptation measures were proposed and discussed with local stakeholders with an updated and improved data set. On that basis, the first set of recommendations for climate change adaptation measures at local level was drafted and circulated among stakeholders in the final workshop of the VIETADAPT project.

Participants also discussed about further cooperation in the second phase of the project in order to conduct a feasibility study on the proposed climate change adaptation strategy. The study shall also have analysis on cost-benefit and uncertainty with the support of scenario workshops.

Finally, it could be said that any new project that attempts to use this approach needs a critical process—building trust between stakeholders. Once stakeholders see that their ideas are welcomed and appreciated, they would make positive contributions. It is obvious that information and experiences from local stakeholders play a critical role in implementing the workshop results, for example the planning process at local level.

References

Bedstedt B, Gram S (2013) Participatory climate change adaptation in Kalundborg, Denmark. In: Schmidt-Thomé P, Klein J (eds) Climate change adaptation in practice—from strategy development to implementation. Wiley Blackwell, London

IPCC (2012) Managing the risks of extreme events and disasters to advance climate change adaptation. A special report of working groups I and II of the intergovernmental panel on climate change, Cambridge University Press, Cambridge

Petersell V, Suuroja S, All T, Shtokalenko M (2013) Impacts of sea level change to the West Estonian coastal zone towards the end of the 21st century. In: Schmidt-Thomé P, Klein J (eds) Climate change adaptation in practice—from strategy development to implementation. Wiley Blackwell, London

Rimkus E, Kažys J, Stonevičius E, Valiuškevičius M (2013) Adaptation to climate change in the Smeltalė River basin, Lithuania. In: Schmidt-Thomé P, Klein J (eds) Climate change adaptation in practice—from strategy development to implementation. Wiley Blackwell, London

Schmidt-Thomé P, Klein J (eds) (2013) Climate change adaptation in practice—from strategy development to implementation. Wiley Blackwell, London, 327p

Slocum N (2003) Participatory methods toolkit. A practitioner's manual. Available under http://www.kbs-frb.be/uploadedFiles/KBS-FRB/Files/EN/PUB_1540_Participatoty_toolkit_New_edition.pdf. Accessed 28 Nov 12

Wollenberg E, Edmunds D, Buck L (2000) Using scenarios to make decisions about the future: anticipatory learning for the adaptive co-management of community forests. Landscape Urban Plann 47:1–2

Chapter 6
Climate Change Adaptation Measures

Abstract As a result of training of young scientists in Finland and Vietnam, scientific field investigations, environmental and climate change impact modeling and the scenario workshops with local stakeholders first climate change adaptation measures were developed for the two cases studies in Vietnam. The adaptation measures have a sectoral approach and are according to timescales of implementation. The adaptation measures are funded mainly by national budgets of the respective research institutions and environmental authorities.

6.1 Climate Change Adaptation Measures in Thanh Hoa Province

Thanh Hoa is currently setting up an action plan to respond to climate change. The impact assessment of climate change and sea level rise in different sectors and areas of Thanh Hoa province within the framework of the VIETADAPT project has provided information for proposing climate change response measures. Also, adjustments concerning the socio-economic development address the master plan of the province to ensure sustainable development.

Solutions by sectors
The study showed that water supply is the most concerning issue in Thanh Hoa province. With regard to water resources, the feasible and effective responses include management measures such as:

- Exhaustive management of water resources of the rivers Chu and Ma
- Increasing public awareness of climate change, and safe and appropriate utilization of surface and ground water resources
- Conducting water supply planning and locating groundwater resources

© The Author(s) 2015   79
P. Schmidt-Thomé et al., *Climate Change Adaptation Measures in Vietnam*,
SpringerBriefs in Earth Sciences, DOI 10.1007/978-3-319-12346-2_6

- Creating sustainable water resource development plan in line with the socio-economic development master plan of the province. Especially, the arid areas need to develop facilities to store water for use during the dry season. Households in these areas should have rain-water tanks and local authorities shall support poor families to comply with these recommendations and regulations
- Applying water-saving irrigation, like rotating irrigation method over channels, improving water pumping registration systems, improving field and multi-field dams, reduce waste water, control salinity, supplement/improve and replace pumping stations of Chau Loc, Thieu Xa, Dai Loc, Ba Dinh and Nga Vinh.
- Dredge estuaries and seaports that link to pumping stations of Hoang Khanh, Quang Loc, Cong Phu, Thieu Xa;
- Invest to upgrade and fortify sea dykes and estuary dykes protecting communes in order to prevent flooding as well as soil and water salinisation

From the analysis of the status quo and development plan for agricultural sector, as well as potential impacts of climate change to crop plants and breeding animals a number of solutions were proposed including efficient use of crop land, use of plants that benefit from changes in climate conditions, applying biotechnology in breeding selection to ensure abilities of anti-disease and extreme climate adaptation and, conducting research and selection of appropriate sugarcane varieties and cultivation methods to ensure economic efficiency.

Concerning forestry, illegal logging is a large problem in the Thanh Hoa province, also in protected areas. This leads to an uncontrolled deterioration of forest areas and soil losses. Hence, forest management and development should be enhanced with special focus on protection of primeval forest. The aim is to strengthen the prevention of forest fires and forest pestilent insects, to promote afforestation and forest restoration, to stop deforestation in order to minimize damages to the ecosystem and to increase forest coverage. Especially afforestation in deserted land and hills, and uncultivated land with forest trees (Luong Bamboo—*Dendrocalamus membranaceus* Munro, *Pinus latteri*, *Manglietia conifera*, gum tree, casuarina, cinnamon, rubber), perennial trees and xerophyte fruit trees is recommended. Also grass planting in large areas for feeding cows, goats, sheep etc. in mountainous areas, as well as afforestation and protection of coastal forests are key options in sustainable forest management.

In general, natural hazard forecasting and the dissemination of information shall be enhanced. Therefore networks for environment and natural hazard monitoring and forecasting shall be improved and communal early warning systems for flood and other natural hazards installed.

6.2   Proposal of Priority Projects in the Period 2013–2020

On the ground of researches and questionnaire results, several key impacts from climate change were defined as follows:

- Storm is one of the most frequent and dangerous natural hazard in the Thanh Hoa province. There are 5–7 storms and tropical depressions annually, mostly between August and October. The second most important hazards are floods and flash floods which are most frequent between June and October. Flash floods are a frequent threat to almost 11 mountainous districts of the province. The third essential natural hazard in Thanh Hoa in terms of damages is drought.
- Many areas are highly vulnerable to erosion and sedimentation. Salinisation of aquifers, water bodies and soils is a threat to coastal districts at different levels.

In order to support the implementation of climate change adaptation measures, concrete projects were proposed as a joint effort between scientists and local stakeholders. These have been devised, analyzed and selected based on several criteria such as urgency, inter-disciplinarity, feasibility and sustainability. The proposed climate change adaptation measures are summarized in Tables 6.1 and 6.2.

6.3   Climate Change Adaptation Measures in the Ba Ria–Vung Tau Province

Based on the research on impacts of climate change and sea level rise as well as the master plan for socio-economic development of the province, several highly vulnerable areas have been identified in Ba Ria–Vung Tau province. The province has set up and implemented the action plan for responding to climate change and piloted several climate change response measures.

Proposed solutions by sectors
Research results showed that the most concerning issues are coastal and estuary erosion, salinity and natural disaster damages.

For water resources, solutions of short, medium and long term were proposed and analyzed, including technical and non-technical ones. The outstanding ones are review, upgrade and construction of salinisation protection, irrigation and drainage systems. Further, it is important to implement at programs and set up appropriate monitoring system to conduct regular checks over water quality in ponds, lakes, rivers and streams. Plans for exploitation, utilization, management and protection of groundwater and surface water in line with policies for management of water resources in river basin and adapting with climate change must be further developed. Mechanism for extensive management of water resources over the entire river basins, meeting demands of water users must be improved, and policies for encouraging and raising awareness of water saving among the community shall be implemented.

Table 6.1 Some projects that need prioritization regarding to climate change adaptation measures in the Thanh Hoa province

No.	Project name/action	Content	Time of implementation	Cost estimation (million VND)	Owner
Priority projects in the period from 2013 to 2015					
1	Developing and upgrading monitoring and warning systems for floods and flash floods	Developing and upgrading monitoring and warning systems for floods and flash flood in 11 mountainous areas	2013–2015	11.000	Provincial hydro-meteorological center
2	Development of natural disaster early warning maps and other documentations guiding the preparation and response to natural disasters	Natural hazard warning maps	2013–2015	1.400	Provincial department for agriculture and rural development
		Documentation guiding the preparation and response to disaster consequences; organization of training courses on preparation and response to disaster consequences in each community area			
3	Assessment of erosion in the province and proposal of solutions	Assessment of erosion in the province and proposal for mitigation	2013–2015	1.500	Provincial department for science and technology
4	Upgrading of marine and riverine levees	Further Investment for upgrading of existing marine and riverine levees in districts: Nga Son, Hau Loc, Hoang Hoa, Quang Xuong, Tinh Gia, Sam Son town	2013–2015	360.000	Provincial department for agriculture and rural development
		Investigations to build new marine and riverine levees in areas that are vulnerable to flood, erosion and sea level rise in the future			

(continued)

Table 6.1 (continued)

No.	Project name/action	Content	Time of implementation	Cost estimation (million VND)	Owner
5	Promoting education and communication for awareness rising on climate change	Development of programs to raise awareness on environment protection and climate change among the communities, public servants and unions, organization over the entire province	2013–2015	1.200	Provincial departments for natural resources and environment; culture, sport and tourism
		Development of adaptation plans			
6	Revision of the land use plan to adapt with climate change and sea level rise	Revision of the land use plan until the year 2020 to adapt to climate change and sea level rise—orientation for land use until 2030 and 2050	2013–2015	1.000	Provincial department for natural resources and environment
Priority projects in the period from 2015 to 2020					
1	Construction of storm shelter anchorage system for fishing-boats	Construction of storm shelter anchorage system for fishing-boats in De dyke (Hau Loc), Sao Sa, Choan dyke (Nga Son), Quang Thach (Quang Xuong), Lach Bang (Tinh Gia)	2015–2020	173.000	Provincial department for agriculture and rural development
2	Reforestation of mountainous and coastal districts	Reforestation of treeless lands and hills to prevent erosion and desertification	2015–2020	100.000	Provincial department for agriculture and rural development
		Reforestation of mangroves, river mouths and bays			
3	Flood protection from four key rivers in Thanh Hoa province	Planning of flood protection systems for the four key rivers in the Thanh Hoa province	2015–2020	9.000	Provincial department for natural resources and environment

Table 6.2 Priority projects for groundwater resources in the Thanh Hoa province (*million VND*)

No.	Project	2013–2015	2015–2020	Estimated cost
1	Development of the network for ground water resource monitoring		10,000	Provincial budget and donors from international organizations
2	Development of a (complete) model and evaluation of modelled climate change impacts on groundwater resources under different models	5,000		
3	Groundwater recharge investigation		4,000	
4	Preparation of studies on groundwater protection		6,000	
5	Planning the use of groundwater (exploitation, protection and prevention of contamination)	1,000	10,000	
	Total	6,000	30,000	

In agriculture areas, key solutions comprise revision and adjustment of plans for agriculture production in ecological areas, improvement of end-to-end production processes from agriculture production, foodstuff processing, breeding processes and waste management to both safeguard the environment and adapt to climate change, implementation of programs and projects for improving forest quality and forest protecting capabilities, especially plantation of upstream protective forests and mangroves, coastal mangroves as protection of sands and waves, conduction of surveys and assessments over marine ecosystems (coral reef, sea grass etc.), rare animals to prepare the ground for development of restoration models and expansion of mangroves, protecting the ecosystem and aquatic animals that taking into account climate change impacts. Solutions for sustainable protection and development of fishery industry that incorporate fishery environment protection comprise the development of production procedures in line with restoring ecosystem and protecting aquatic resources.

6.4 Proposal of Priority Projects in the Period 2013–2020

On the ground of researches and questionnaire results, several key impacts from climate change were defined as follow:

- The areas that are most vulnerable to erosion and deposition are Cua Lap in Vung Tau City and Phuoc Tinh in Long Dien District; Loc An in Dat Do District, Ho Tram, Ho Coc and Binh Chau in Xuyen Moc District.

Table 6.3 Some priority projects regarding climate change adaptation measures in the Ba Ria–Vung Tau province

No.	Project name/action	Content	Time of implementation	Cost estimation (million VND)	Owner
Priority projects for the period 2013–2015					
1	Assessment of erosion in the province and proposals for solutions	Assessment of erosion in marine and riverine coastal areas in the provinces and proposals for countermeasures	2013–2015	600	Provincial department for science and technology
2	Assessment of climate change and sea level rise impacts on mangrove ecosystems of Ba Ria–Vung Tau and develop response measure	Assessment od climate change impacts on mangrove system	2013–2015	500	Provincial department for agriculture and rural development
		Proposal of measures for rehabilitation and protection of coastal mangroves and wetlands for protection from tsunami, floods, storms and coastal erosion			
3	Revision of the land use plan to adapt to climate change and sea level rise	Revision of the land use plan until the year 2020 to adapt it to climate change and sea level rise—orientation for land use until 2030, 2050	2013–2015	1.000	Provincial department for natural resources and environment
4	Upgrading of water reservoirs in the province	Conduction of surveys and evaluation of management capabilities of the water provision capacity over all reservoirs. Assessments on possibilities to increase reservoirs' volume. Linking of reservoirs to regulate water discharge to store water in dry seasons and retain floods in rainy seasons	2013–2015	16.000	Provincial department for agriculture and rural development

(continued)

Table 6.3 (continued)

No.	Project name/action	Content	Time of implementation	Cost estimation (million VND)	Owner
5	Upgrading of marine and riverine levee systems	Investment for upgrading of existing sea and river levees	2013–2015	739.934	Provincial department for agriculture and rural development
	Construction of new levee in Hai Dang				
	Seaside protection construction in Ben Loi, Loc An				
	Upgrading Chu Hai levee				
	Repair and upgrade Phuoc Tinh marine levee, phase II	Investment to build up new river and marine levees in areas that are vulnerable to flood, erosion, sea level rise in the future			
6	Preparation of documentations guiding the preparation and response to natural disaster consequences for each community	Preparation of maps for natural disaster early warning and documentations guiding the preparation and response actions, organization of trainings and coaching	2013–2015	620	Provincial department for agriculture and rural development
7	Raising of climate change awareness for all public servants, communities, organizations and unions of Ba Ria–Vung Tau province	Organization of training programs	2013–2015	1.000	Provincial department for natural resources and environment
		Compilation of communication documents on climate change			
		Building of websites on climate change			

(continued)

Table 6.3 (continued)

No.	Project name/action	Content	Time of implementation	Cost estimation (million VND)	Owner
Priority project for the period 2016–2020					
1	Enhancement of climate change adaptability of coastal tourism; proposals for adaptation	Assessment of climate change impacts on coastal tourism activities	2016–2020	2.000	Provincial department for culture, sport and tourism
		Development of adaptation plans			
2	Upgrading hydro-meteorology monitoring, improvement of forecasting capabilities	Organization and enhancement of existing systems	2016–2020	20.000	Provincial hydro-meteorological center
		Enhancement of capabilities of assessment and analysis of monitoring data			
		Exchange of data with the relevant Institutes; incorporation of information and data from foreign centers			
3	Upgrading of marine and riverine levees	Further Investment for upgrading existing sea and river levees	2016–2020	600.000	Provincial department for agriculture and rural development
	Seaside protection constructions in Ho Tram, Ho Coc, Loc An 2, Phuong Lam—Lo Voi, Tan Phuong—Phuong Binh, Phuong Hung 1, Phuong Hai 1	Investigations to build up new riverine and marine levees in areas that are vulnerable to flood, erosion and sea level rise in the future			
4	Implementation of pilot models for natural disaster mitigation and climate change adaptation	Implementation of pilot models	2016–2020	600	Provincial department for agriculture and rural development
		Organization of site visit at pilot models			
		Information workshops			

Table 6.4 List of priority projects for groundwater resources in the Ba Ria–Vung Tau province

No.	Project name/action	Goals	Content	Deliverables	Budget (million VND)	Source of budget and time for implementation
1	Groundwater mapping	Search and evaluation of groundwater resources in Long Dat—Xuyen Moc	Assessment of groundwater reserves in coastal areas within depth of 5 m	Groundwater potential assessments	4,000	Local budget Time: 2014
				Geological mapping at 1/100,000 scale following geophysical materials		
2	Groundwater resource assessment	Preliminary assessment of groundwater at the scale 1/50,000 for the entire province	Explicit assessment of groundwater potential reserves as basis for revision of future use of groundwater resources	Hydro-geologic mapping at the scale 1/50,000 to meet national standards	10,000	Local budget or request for support from the government Time: 2014–2015
				Groundwater resources map		
3	Assessment of groundwater pollution caused by socio-economic development	Determination of pollution sources, forecast of evolvement of pollution and proposal of remediation measures	Data collection	Groundwater pollution maps	8,000	Local budget Time: 2015–2016
			Conduction of supplemental surveys			
			Collection and analysis of samples			
			Groundwater modelling			
			etc.	Forecasting, early warning and remediation measures		

(continued)

Table 6.4 (continued)

No.	Project name/action	Goals	Content	Deliverables	Budget (million VND)	Source of budget and time for implementation
4	Assessment of groundwater reserve changes under impacts of climate change and sea level rise	Assessment of groundwater potential under scenarios of climate change and sea level rise	Data collection Permeability tests Monitoring, VES Modeling etc.	Groundwater resources map at the scale of 1/50.000 Climate change scenario impact evaluations	5,000	Local budget Time: 2016–2018
5	Research and proposition of optimum exploitation solutions for catchment areas	Determination of optimum exploitable reserve and recommendation for well fields	Data collection Conduction of supplemental surveys Groundwater modelling etc.	Proposal for appropriate exploitation of groundwater for catchment areas	3,500	Local budget and private sector investments Time: 2016–2018
6	Research on potential for artificial groundwater recharge in urban areas by collected rain water	Determination of implementation potential for artificial recharge for main aquifers	Construction of pilot facilities in 3 areas (wells, tanks…) Monitoring of water level and quality Modeling	Guidelines for supplementing groundwater by rain-water Proposal of design	4,000	Local budget Time: 2016

(continued)

Table 6.4 (continued)

No.	Project name/action	Goals	Content	Deliverables	Budget (million VND)	Source of budget and time for implementation
7	Study and assessment of impacts of mineral exploitation on groundwater quality and quantity	Determination and study of mineral exploitation activities that affect groundwater resources	Data collection, Supplementary investigations	Forecast of key impacts	7,500	Local budget and enterprises Time: 2018–2020
			Water sampling and analysis			
			Dynamic groundwater monitoring			
			Modeling	Proposal of appropriate management of mineral resource exploitation		

- In the Vung Tau province, water resources are increasingly scarce, weather extremes are increasing and ecosystems are deteriorating. The province has limited water resources, with only two main rivers providing fresh water being Dinh and Ray River. Climate change causes changes to precipitation patterns, increases in temperature, and potentially leads to water scarcity during dry seasons. Salinisation and groundwater pollution are increasing threats. For example groundwater wells in Ba Ria and Tan Thanh are the key sources of fresh water are highly vulnerable to salinisation.

From general solutions, specific ones have been devised, analyzed and selected base on criterias similar to those of Thanh Hoa (see above). Concrete proposals for projects to implement climate change adaptation measures are set out in Tables 6.3 and 6.4.

Chapter 7
Conclusions and Recommendations

Impacts of socio-economic development and climate change have been evaluated on groundwater and surface water resources, as well as some environmental factors for the two coastal provinces of Thanh Hoa and Ba Ria–Vung Tau.

An overview over the current status and vulnerabilities of groundwater and surface water resources were developed and analysed with stakeholders. Socio-economic development and climate change impacts on various areas and sectors of the two provinces were conducted in close cooperation between scientist and local stakeholders.

The VIETADAPT project has undertaken a detailed research and evaluation of impacts of climate change on various elements of temperature, rainfall and sea level rise on the basis of baseline data and climate change scenarios. The project also included the evaluation of impacts of climate change on flow, accumulation and erosion at estuaries and saline intrusion towards surface and groundwater in coastal regions of two provinces of Thanh Hoa and Ba Ria–Vung Tau.

A holistic overview on current and emerging vulnerabilities on water resources, taking both climate change and socio-economic development impacts into account was developed. These vulnerabilities were elaborated in close science-stakeholder cooperation.

Activities included in the VIETADAPT project have made positive contribution towards the strengthening of national capacity on adaptation to climate change in the two coastal provinces of Thanh Hoa and Ba Ria–Vung Tau.

With the support of the scenario workshops, the outputs from the BaltCICA project (Schmidt-Thomé and Klein 2013) have proven to be a useful tool in the communication between scientists and local stakeholders.

The achieved outputs of the project provide necessary information in support of localities in developing and implementing activities for adaptation to climate change; at the same time contributing endorsing adjustment, development and integration of climate change issues into future programs, plans, master plans; and finally the enhancement of abilities of local residents as well as key bodies in developing climate change adaptation measures.

In the framework of the project, the evaluation of impacts of climate change on mangrove forests, alluvial grounds and coastal isles have not been included.

© The Author(s) 2015
P. Schmidt-Thomé et al., *Climate Change Adaptation Measures in Vietnam*,
SpringerBriefs in Earth Sciences, DOI 10.1007/978-3-319-12346-2_7

Therefore, it is necessary to consider the evaluation of impacts of climate change on these areas and the vulnerability of these areas as a result of climate change.

Based on the project's activities, it is recommended that the management bodies/agencies, policy makers need to closely coordinate with the local people to carry out comprehensive assessments to develop sustainable solutions for adaptation to climate change. It is necessary to organize training courses to enhance the knowledge on vulnerabilities, impacts of both socio-economic developments and climate change, and put forward appropriate measures to ensure that the project's activities are applicable effectively in practice.

In order to strengthen the capacity to adapt to climate change and achieve sustainable socio-economic development, there needs to be close coordination between stakeholders under the programs relating to climate change, as well as concerned agencies of Ministries/sectors and localities.

The results achieved by this project were made possible by the financing agency, the Finnish Ministry of Foreign Affairs and its ICI instrument; the excellent cooperation between the implementation agencies GTK, IMHEN, and NAWAPI; and the active participation from officers of DONRE and the Department of Science and Technology of the two provinces of Thanh Hoa and Ba Ria–Vung Tau.

Reference

Schmidt-Thomé P, Klein J (eds) (2013) Climate change adaptation in practice—from strategy development to implementation. Wiley Blackwell, London, 327p

Appendix
Survey Questionnaire for the Project Develop and Implement Climate Change Adaptation Measures in the Coastal Areas of Vietnam

A.1 General Information

Interviewee ....

Date:...

Organization: ....

Department: ....

Address:..

Phone ....

Email:...

© The Author(s) 2015
P. Schmidt-Thomé et al., *Climate Change Adaptation Measures in Vietnam*,
SpringerBriefs in Earth Sciences, DOI 10.1007/978-3-319-12346-2

A.2 Awareness of Climate Change Impacts

A.2.1 Which Natural Hazards (Disasters) are Caused by Climate Change in Your Areas?

Scoring (if available): 5 = Highly important, need special attention; 4 = important, need attention; 3 = important; 2 = Yes, not very clear; 1 = Not important in the area.

Natural danger (disaster)	Yes	No	Score
Tornado			
Flood			
Saline ground water.			
Saline surface water.			
Saline land			
Sea Level Rise			
Landslide			
Storm surge			
Heavy rain			
Drought			
High temperature			
Other (please specify):			

A.2.2 What are the Key Vulnerabilities Caused by Climate Change in Your Area ?

Scoring (if available): 5 = Highly important, need special attention; 4 = important, need attention; 3 = important; 2 = Yes, not very clear; 1 = Not important in the area.

Society	Yes	No	Score
Houses			
Schools, hospitals, fire-fighting stations etc			
Health problems (diseases, quantity and quality of drinking water etc)			
Information, news, response measures			
Vulnerable group such as the elderly, children			
Family livelihood			
Other (please specify):			

Economic	Yes	No	Score
Change in economic structure			
Change in land use structure (Decrease in land area for agriculture, aquaculture; Increase in land for industrial production, tourism, urban)			
Infrastructure: Roads, sluices, electricity cables etc			
Other (please specify):			

Environment	Yes	No	Score
Shortage of ground water: quantity and quality of water			
Shortage of surface water: quantity and quality of water			
Saline land			
Land erosion			
Land deposition			
Contamination and pollution due to human's activities			
Loss of biodiversity, deteriorated nature			
Loss of agriculture land and forest.			
Landslide			
Other (please specify):			

A.3 Response Measures for Current Climate Change

A.3.1 Response Measures for Current Climate Change ?

Scoring (if available): 5 = Highly important, need special attention; 4 = important, need attention; 3 = important; 2 = Yes, not very clear; 1 = Not important in the area.

Climate change response measures	Yes	No	Score
Flood or tornado early warning systems at residential areas.			
Salinity early warning system			
Sea dyke to confront sea level rise			
Seashore erosion protection			
Ensure the provision of drinking water, or conduct regular suction for salinity confronting (long term adaptation) or due to dangerous			
Educate and raise awareness of policy-makers and stakeholders			
Educate and raise awareness of local people			
Cooperation between stakeholders of different beneficial groups			
Readiness of hospitals, fire-fighting forces etc in dangerous situations			
Other (please specify):			

A.3.2 How Does the Land Use Plan Take Climate Change Impacts into Consideration in Your Areas?

Scoring (if available): 5 = Highly important, need special attention; 4 = important, need attention; 3 = important; 2 = Yes, not very clear; 1 = Not important in the area.

Urbanization	Yes	No	Score
Analysis of water balance			
Housing conditions (environment, structure, durability)			
Infrastructure Efficiency:			
Wastewater treatment			
Wastewater management			
Preparation for urgent cases			
Other (please specify):			
Industrialization	**Yes**	**No**	**Score**
Analysis of water balance			
Environmental impact assessment			
Environment monitoring			
Other (please specify):			
Tourism	**Yes**	**No**	**Score**
Analysis of water balance			
Change in land use, potential conflicts with other economic sectors			
Other (please specify):			
Agriculture	**Yes**	**No**	**Score**
Social impacts due to changes in economic sector			
Change in land use, potential conflicts with other economic sectors (urbanization. tourism)			
Salinity => adaptation measures (dykes, early warning system)			
Other (please specify):			

A.4 Recommendations

A.4.1 What are the Main Concerns in the Near Future in Your Areas Regarding Climate Change Impacts?

....

....

Date….. month….year……
(signature)